1

Kristall und Technik

Akademie-Verlag · Berlin

1. Heft · 5. Jahrgang · 1970

Hinweise für die Autoren

Die Zeitschrift „Kristall und Technik" nimmt folgende Veröffentlichungen auf: Originalarbeiten, Zusammenfassungen, kurze Mitteilungen, Buchbesprechungen, bevorzugt in englischer Sprache,

Alle zum satzreifen Manuskript gehörenden Unterlagen sind beizufügen.

Das Manuskript soll in Maschinenschrift im Format A 4 (21 × 29,7 cm) auf weißem, tintenfestem Papier, einseitig, mit zweizeiligem Abstand der Zeilen und einem linken freien Rand von 3 cm geschrieben sein. Die Manuskriptblätter sollen fortlaufend ab „1" numeriert werden.

Noch nicht für den Druck überarbeitete Vortragsmanuskripte werden in der Regel nicht angenommen.

Der Kopf des Manuskriptes soll folgende Zeilen enthalten: Abgekürzte Vornamen und den Familiennamen des Verfassers, genaue Bezeichnung der Arbeitsstelle des Verfassers, Überschrift der Arbeit.

Der Arbeit ist eine Zusammenfassung in der Original- und einer Fremdsprache (deutsch, englisch, französisch oder russisch) voranzustellen.

Formeln, Zeichen, Buchstaben usw., für die keine Schreibmaschinentypen vorhanden sind, müssen deutlich lesbar mit Tinte geschrieben sein.

Nomenklatur des Chemischen Zentralblattes benutzen.
O als Null ist mit der Schreibmaschine einfach zu unterstreichen, der Großbuchstabe O nicht.

Die Absätze sind deutlich im Manuskript zu kennzeichnen (durch Einrücken von mindestens 3 Anschlägen).

Für die Auszeichnung des Manuskriptes ist zu beachten:

Überschriften im Text und zum Beginn des Beitrages sind übersichtlich anzuordnen und nur durch eine einfache Unterstreichung kenntlich zu machen.

Tabellenüberschriften sind nicht besonders kenntlich zu machen, sie werden normal gesetzt.

Auszeichnungen innerhalb des Textes werden in der Regel von der Redaktion erledigt. Es bedeuten:

hellrot ～～～～～ = Kapitälchen
blau ～～～～～ = kursiv

hellrot | = Kleindruck (senkrechter roter Strich am linken Manuskriptrand)

Tabellen sind gesondert als Manuskript zu schreiben und am Schluß des Beitrages einzuordnen.

Fußnoten sind auf gesondertem Blatt fortlaufend numeriert aufzuführen.

Abbildungen [„Fig." (= Figur)] sind als gut reproduzierbare Vorlagen dem Manuskript in gesondertem Umschlag beizulegen.

Fotos sind auf weißem Hochglanzpapier und im Mindestformat von 6 × 6 cm zu liefern. Bildausschnitte und eventuell noch vorzunehmende Beschriftungen sind auf einem auf dem Foto angebrachtem Deckblatt aus durchsichtigem Material zu vermerken. Die Größen aller Fotos und deren Verkleinerungsmaßstäbe sind nach Möglichkeit einheitlich zu halten.

Alles Nebensächliche bei Halbtonbildern bitte weglassen und von einer Vergrößerung der Vorlagen absehen.

Strichzeichnungen genügen als gut lesbare Skizzen.

Auf der Rückseite der Abbildungsvorlagen sind anzugeben: Nummer der Figur, Name des Autors bzw. der Autoren, abgekürzter Beitragstitel, gewünschte Größe der Figur im Druck in Zehntel-Angaben, z. B. 9/10, 8/10, 5/10 usw.

Bildunterschriften sind auf einem gesondertem Blatt aufzuführen.

Die Literatur soll ohne Bezifferung am Ende der Arbeit in alphabetischer Reihenfolge der Autorennamen aufgeführt werden. Im Text zitierte Autorennamen gelten als Literaturhinweis.

Die Literaturzitate sollen folgende Angaben enthalten:

bei Büchern z. B.: RAST, K.: Physikalisch-chemische Rechnungen, Berlin 1958
bei Zeitschriften z. B.: KRAUSE, H. N., MÜLLER, K.: Kristall u. Technik **1**, 70 (1966)

Für jede veröffentlichte Arbeit werden 50 Sonderdrucke vom Verlag kostenlos abgegeben. Sind mehrere Autoren gemeinsam an einem Beitrag beteiligt, so erhalten sie insgesamt 50 Sonderdrucke kostenlos.

Die Bestellung weiterer Sonderdrucke gegen Berechnung, jedoch nur bis zu einer Anzahl von 200 Stück, muß mit der Einsendung des Manuskriptes der Hauptschriftleitung bekanntgegeben werden.

1

Kristall und Technik

Zeitschrift für experimentelle und technische Kristallographie

BAND 5 · HEFT 1
1970

Akademie-Verlag · Berlin

Herausgeber: Professor Dr. WILL KLEBER, 104 Berlin, Invalidenstraße 43 bzw. 1178 Berlin(-Wendenschloß), Ostendorfstraße 15, und Professor Dr. HERMANN NEELS, 703 Leipzig, Scharnhorststraße 20 bzw. 7254 Machern, Zeititzer Weg 25.
Hauptschriftleiter und verantwortlich für den Inhalt: Professor Dr. HERMANN NEELS, 703 Leipzig, Scharnhorststraße 20. Fernruf: 3 25 19.
Verlag: Akademie-Verlag GmbH, 108 Berlin, Leipziger Straße 3 – 4,
Fernruf: 22 04 41, Telex-Nr. 112 020, Postscheckkonto: Berlin 350 31.
Die Zeitschrift „Kristall und Technik" erscheint vierteljährlich. Bestellnummer: 1082/5/1.
Preis je Heft 20,– M
Gesamtherstellung: VEB Druckerei „Thomas Müntzer", 582 Bad Langensalza
Veröffentlicht unter der Lizenznummer 1526 des Presseamtes beim Vorsitzenden des Ministerrates der Deutschen Demokratischen Republik.

Inhalt

Übersichten

Originalbeiträge

Kurze Originalmitteilungen

Kristall und Technik	5	1	1970	5–49

K. MEYER, D. OBRIKAT, M. ROSSBERG

Deutsche Akademie der Wissenschaften zu Berlin
Zentralinstitut für Physikalische Chemie

Progress in Triboluminescence of Alkali Halides and Doped Zinc Sulphides (I)

Contents

[1]) in Heft 2/1970

1. Definition of Triboluminescence (TL)

Research on triboluminescence has taken a rapid development during the last decade. For this reason it seems to be quite practical to give a summary on its present state.

The term triboluminescence was introduced by WIEDEMANN and SCHMIDT in 1895. According to their interpretation it is the light emission occuring during the mechanical treatment (rubbing, crushing, pressing) of solids which is not in a thermal equilibrium with the radiant system and therefore possibly might show a higher radiation intensity than the intensity of equifrequent radiation of a black body of the same temperature. The mechanical work must be the energy-supplying process, and thermal radiation is to be excluded.

In the light of the latest findings on elementary mechanisms of light emission and the relationship with other crystalline properties, we are faced with the necessity of subjecting the classical phenomenological definition to a critical investigation. The mechanical treatment of crystals generally constitutes a very rough interference. A great number of physical processes may occur within very short time intervals which excite or stimulate the process of photon emission. Among these are: elastic and plastic deformations as well as fracture processes; local and extremely high pressure and temperature peaks; short-period exposures of extremely clean surfaces and the development of free radicals; electric processes, e. g. the development of electric potential differences due to contact potential differences, by the development of electrically charged fracture planes or by the movement of charged dislocations. In addition, the crystal cannot be considered separately, but the material environment, in particular the kind and pressure of gas must be included in the system.

The question for the elementary causes of TL, therefore, reduces to the question as to how the energy added to the system dissipates and which simultaneously passing processes of physical nature have a dominating influence on the emission process.

Comparing the term of TL with other kinds of luminescence, perhaps with photoluminescence (PL), we are faced already with difficulties which indicate the complicated process of the mechanical light excitation. The prefix of common kinds of luminescence always expresses the last stage of excitation. At PL, for example, a crystal is exited to light emission directly, i. e. a direct excitation of the phosphor by the irradiated light takes place. By transferring the terminology to TL, the mechanical energy added to a crystal would have to be transferred directly to the atoms of the solid, e. g. by modified lattice vibrations, and excite the atoms to luminescence. This mechanism is conceivable, it is true, but it has never been identified yet. The mechanical energy is rather transformed into other forms of energy which then constitute the last stage of excitation.

In the light of these facts, we propose to use the term "tribo-induced luminescence" in the most general case if light is emitted during the mechanical treatment of a crystal and if the actual mechanism of excitation is unknown. If the specific process of excitation is known, the following tribo-induced kinds of luminescence might be differentiated:

Triboluminescence (TL)

The excitation of lattice atoms is effected directly by the mechanical energy added, e. g. by way of forced lattice vibrations.

Tribo-induced Chemiluminescence (CL)

The photon emission comes about by chemical processes on the mechanically stressed crystal, e. g. by recombination of broken bonds or radicals (radical luminescence), in particular with the destruction of polymers or by adsorption of gases on activated crystal surfaces (adsorboluminescence).

Tribo-induced Electroluminescence (EL)

The excitation of solid state or ambient and/or adsorbed gas atoms is effected by electric fields which develop themselves during the mechanical treatment. Electric potential differences might occur, e. g., at the elastic deformation of piezo-electric crystals, at the movement of electrically charged dislocations, at the development of electrically charged fracture areas (e. g. (111)-surfaces of alkali halides), by separation of directed defect bonds (e. g. in As_2O_3), or by differences in the contact potentials.

Tribo-induced Radioluminescence (RL)

The process of treatment might cause energetic electrons to be formed (e. g. MEYER, LINKE; LINKE) which upon their impinging on the crystal excite the latter to radiation.

Tribo-induced Photoluminescence (PL)

The tribo-induced luminescence developed due to one of the said excitation mechanisms for its part might generate a photoluminescence due to their action on the crystal. For instance, the emission of a crystal in the long-waves lenght-region (e. g. Mn-activated ZnS) would be conceivable due to the short-wave portion of the spectrum of a preceding gas discharge.

Extending the historic definition one must give due attention to a thermal radiation which was pointed out by Meyer, Obrikat, Rossberg and which will occur especially under conditions of intensive stress. It would have to be referred to as tribo-induced thermal radiation. Figure 1 shows a set-up of the various kinds of tribo-induced luminescences.

The variety of the phenomena described shows clearly how complicated the explanation of the elementary causes of TL is and that the phenomenon is closely linked with the physical processes which go along with a mechanical stress.

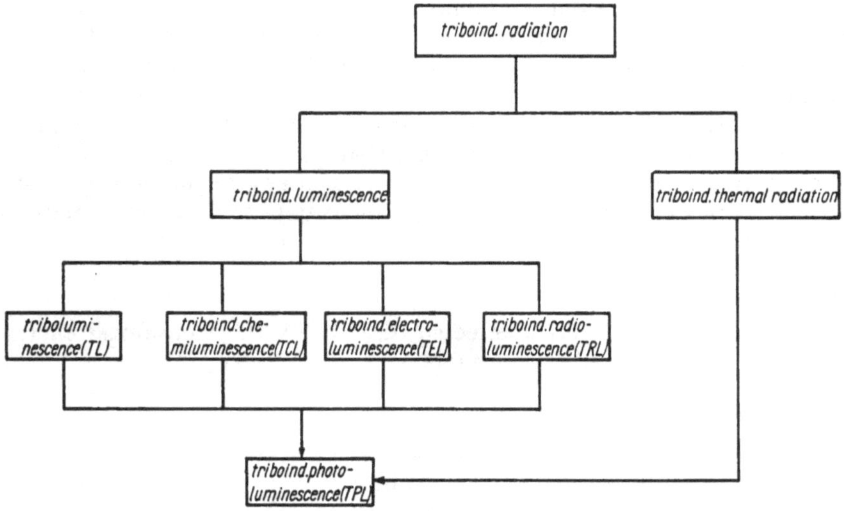

Fig. 1. Scheme of possible kinds of triboinduced radiation

2. General Survey of Previous Studies

The numerous studies on triboluminescent crystals on hand and the fact that the emissivity depends largely on the conditions of treatment make it advisable for classification to be made on the basis of substance groups and methods of treatment. In this chapter we shall describe general relationships on the TL of crystals, whereas the special problems of alkali halides and zinc sulphides will be dealt with in the main part.

2.1 Triboluminescent Crystals

The luminescence property of certain crystals has been mentioned very early already, among others that of quartz, rocksalt (MUSCHENBROCK). In later years, the phenomenon was described repeatedly in the dressing of natural mineral deposits. In the systematic investigation of the TL it was discovered that the light emission is observed both by inorganic and organic substances and, as a rule, is not bound to a definite structure or symmetry of a crystal. We know from experience, however, that crystals without centre of symmetry are more tribo-luminescence-active than those of non-polar character. Therefore, piezo-, pyro- and ferroelectric crystals preferably show triboluminescence. WOLFF, GROSS and STRANSKI have compiled a list of tribo-luminescent crystals, arranged on the basis of substance classes. As the mechanism of light emission is bound to the development of electric fields in the crystal or on its surface, the effect so far has been found on insulators and semi-conductors, but not on pure metals. The TL of shock-treated aluminium investigated by COLEBURN, SOLOW, WILEY has turned out to be a property of the oxide film. Although alkali halides and doped zinc sulphides constitute the model substances in the TL research and these materials have supplied most of the experimental results, we know a series of substances which were tested more intensively for their particular emissivity under different aspects. They include cane sugar, quartz, fluorspar (GROSS, STRANSKI, WOLFF; INOUE; MEYER, POLLY; LONGCHAMBON; BURKE; BEALS; NELSON), mica (OBREIMOFF), semi-conductor elements (JENNY), Mn-activated Na_2ZnGeO_4 (LOBACHEW, KUSMINA, MARTYSHEW), arsenolith (STRANSKI, STRAUSS, WOLFF), tartaric acid (INOUE), Salicylamide (BERSIS, NIKOKAOVURAS).

2.2 Methods of Treatment (mechanical excitation)

TL largely depends on the kind of mechanical treatment and its intensity. Most of the methods used hitherto for the excitation of TL observed the luminous phenomenon by summation of numerous single processes, which are closely superposed as to time and location, in collective crushing which exists, for example in pulverizing, grinding, or sandblasting. The influences of numerous parameters of the mechanical treatment cannot simply be found separately from each other. By the development of electronic measurement techniques, the detection sensitivity has been improved considerably, primary light flashes registered, and mechanical influences on the structure of solids refined and linked up with the change of other physical processes proceeding simultaneously with the luminous process. Different methods were used depending on the intensity of emission and the problem:

Impact stress (KÜRTEN, RUMPF; LONGCHAMBON), collective crushing in the mill (RUMPF; OBRIKAT, MEYER, POLLY), dilatations and compressions of single crystals (BELYAYEW, MARTYSHEW, NABATOW), scratching (BELYAYEW, MARTYSHEW) needle impression (MEYER, POLLY; POLLY, OBRIKAT, MEYER), milling and cutting (MEYER, OBRIKAT, ROSSBERG).

2.3 Parameters for the Identification of TL

The outstanding parameter for describing the TL phenomenon and for its interpretation is the determination of the spectral distribution of the emitted light. Due to the low intensity and the pulsive emission of the photons it was not possible for a long time to obtain plottable spectra. They were obtained only in recent times. The problem is to compare these tribospectra with other emission spectra of the respective substance, possibly with photoluminescence, electroluminescence, fluorescence, or gas discharge spectra. This allows to draw important conclusions concerning the excitation mechanism.

Another parameter is the duration of the light flashes. This quantity, however, cannot be considered as a constant quantity and as typical for a particular substance, it depends largely on the conditions of treatment. The classification of the triboluminescent substances into A-types of short luminescence period ($>10^{-4}$ s) and B-types of a luminescence period $>10^{-3}$ s proposed by WOLFF, GROSS, STRANSKI therefore seems to be too superficial.

It has been realized in recent years that TL frequently is linked up with other physical processes occuring in the crystal or on its boundary surface and that it is of major importance to cover the same and to relate them to the light emission. TL-research, therefore, is closely linked up with the investigation of physical changes of a crystal in the course of its mechanical treatment. This includes first of all the following parameters: elastic and plastic deformations, fracture origin, production of potential differences, strain rate, conductivity, release of gases. The example of the alkali halides and zinc sulphides is to serve as a quantitative representation for a few relationships.

3. Zinc Sulphides

3.1 Results

3.1.1 Spectral Distribution of the TL-Radiation

A determinant characteristic of each luminescence of a luminophor is the spectral distribution of the luminescent emission. It permits a direct statement on the energy and energy distribution of the electron transitions and a restriction of the mechanism by furnishing the conditions necessary for the process of luminescence. The most important step in the treatment of the TL of the four best-known doped zinc sulphides therefore must be to acquire the TL emission spectra. They permitted

a. the splitting of the total emission in a gas discharge portion and a portion of the solid-state luminescence to be determined, because sharp lines — e. g. nitrogen lines — or broad luminescent bands would have to occur under these circumstances.

b. moreover, a comparison of the tribo and photoluminescence (PL)-spectrum furnished an answer to the question if the same centres are effective in both kinds of luminescence, and

c. the hypothesis that the TL of the zinc sulphides may be attributed to an electroluminescence (EL) to be checked.

The results described and discussed below were obtained from powdered zinc sulphides of a maximum grain size of approx. 50 μm. A complete summary of important spectral data is given in Table 1.

3.1.1.1 TL-Spectrum of ZnS: Mn

As early as 1916, Waggoner, using a spectrometer, visually observed the relatively strong TL-emission of ZnS:Mn as compared with other zinc sulphides. He found an agreement with the luminescence excited by X-rays. Waggoner used a rotating circular disk for the generation of light, the surface of which had a ZnS:Mn-layer pasted on it, and a steel brush rubbing on that layer. The light flux obtained by him in combination with the spectrometer used turned out to be insufficient as to be able to register the spectrum objectively.

Nelson succeeded in obtaining the first photography of the TL-spectrum of ZnS:Mn in 1926. He investigated four natural zinc blende samples (sphalerite) and one artificially made ZnS:Mn-phosphor by means of the experimental setup described by Waggoner. The TL and PL band maxima were found to be at 575 nm each with the last sample and at 590 nm and 575 nm respectively with the four first samples. These results were confirmed by Chudáček; Schönert, Kürten; Obrikat, Meyer, Polly in 1967. In another study, Meyer, Obrikat were able to prove that the crystal structure (sphalerite or wurtzite), the quantity concentration in the range between $5 \cdot 10^{-4}$ g — $5 \cdot 10^{-2}$ g Mn/g ZnS and secondary dopings of Cu and Ag between 10^{-4} and 10^{-6} g/g ZnS at a constant Mn-content have no influence on the TL-spectrum. In agreement with Sodomka (1968), a displacement of the TL-bands, as compared with the PL and EL-bands, in the direction of higher wavelength by an average of 1.5 nm was discovered. This value was obtained by taking the mean of a total of 15 measurements, with 2 and 1 nm occuring as maximum and minimum values respectively of the band displacement. The TL and EL spectra of

Table 1
Comparison of spectral data of TL and PL

phosphor		triboluminescence						photoluminescence					
		blue band		green band		yellow band		blue band		green band		yellow band	
		λ_{max}	Q_{bl}	λ_{max}	Q_{gr}	λ_{max}	Q_{ye}	λ_{max}	Q_{bl}	λ_{max}	Q_{gr}	λ_{max}	Q_{ye}
ZnS:Cu, Cl	10^{-3} g Cu/g ZnS	462	0.56	522	0.32	590	0.12	470	0.10	525	0.82	590	0.08
			0.46		0.43		0.11						
ZnS:Cu, I	10^{-3} g Cu/g ZnS	455	0.79	522	0.21	—	—	461	0.59	525	0.41	—	—
ZnS:Cu, Cl	10^{-4} g Cu/g ZnS	453	0.25	520	0.75	—	—	—	—	525	1	—	—
(ZnCd)S:Cu.Cl	10% Cd	454	0.30	521	0.21	567	0.49	—	—	525	0.14	572.5	0.86
	20% Cd	472	0.21	520	0.12	602	0.67	—	—	525	0.07	609	0.93
	30% Cd	472	0.07	520	0.23	640	0.70	—	—	525	0.10	650	0.90
ZnS:Ag		452	1	—	—	—	—	456	1	—	—	—	—
ZnS:Cl		462	1	—	—	—	—	466	0.83	525	0.17	—	—
ZnS:Mn	$5 \cdot 10^{-2}$ g Mn/g ZnS	—	—	—	—	590	1	—	—	—	—	589	1
	10^{-2} g Mn/g ZnS	—	—	—	—	592	1	—	—	—	—	590	1
	$5 \cdot 10^{-3}$ g Mn/g ZnS	—	—	—	—	593	1	—	—	—	—	590.5	1
	10^{-3} g Mn/g ZnS	—	—	—	—	594	1	—	—	—	—	593	1
	$5 \cdot 10^{-4}$ g Mn/g ZnS	—	—	—	—	591,5	1	—	—	—	—	590	1
ZnS:Mn, Ag	10^{-2} Mn, 10^{-4} Ag	—	—	—	—	591	1	—	—	—	—	590	1
	10^{-5} Ag	—	—	—	—	590.5	1	—	—	—	—	589	1
	10^{-6} Ag	—	—	—	—	592	1	—	—	—	—	590	1
ZnS:Mn, Cu	10^{-4} Cu	—	—	—	—	590.5	1	—	—	—	—	589	1
	10^{-6} Cu	—	—	—	—	590.5	1	—	—	—	—	589	1

$$Q_{bl,\,gr,\,ye} = \frac{J_{bl,\,gr,\,ye}}{J_{bl} + J_{gr} + J_{ye}}$$ = portion of blue, green or yellow band related to the total emission

ZnS: 5 · 10^{-3} Mn are shown on Figure 2, representing all the luminophores investigated.

Contrary to CHUDÁČEK (1967) who found gas discharge lines (2% of the total emission) at 400, 402, 407, and 500 nm, MEYER, OBRIKAT did not find any symptoms of a line TL-emission, although the detection limit was increased

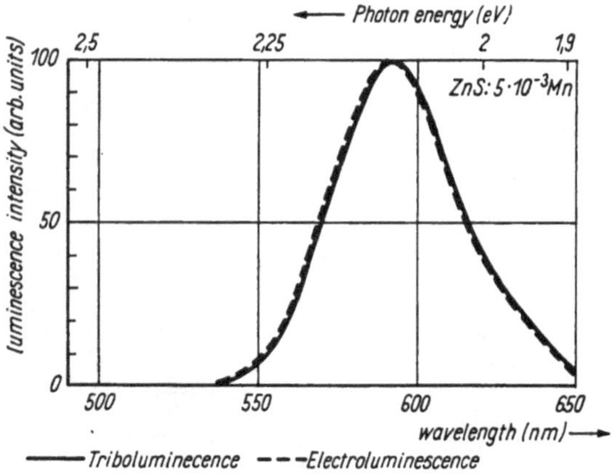

Fig. 2. Comparison of TL- and EL-spectrum of ZnS: 5 · 10^{-3} Mn

to 0.001% of the band maximum at 590 nm. The authors obtained the same result with all other TL-spectra of other zinc sulphides. The experimental results deviating from each other, therefore, may only be explained due to the different mechanical methods of treatment. CHUDÁČEK's rotating drill face-ground at the front and rubbed by the layer probably leads to a considerable contact charging of the contact surfaces; in particular if a contact partner happens to be a good insulator and the contact areas are of microscopic dimensions ($>10^{-4}$ cm²), gas discharge phenomena might occur. Under similar conditions, a gas discharge luminescence, which is to be referred to as frictional luminescence here, was detected by POLLY, OBRIKAT, MEYER. As MEYER, OBRIKAT used a jet mill developed by RUMPF, which provides only very small contact areas, such a frictional luminescence could be excluded. The TL emission spectra thus show that the TL of powdery ZnS:Mn and of all other ZnS-phosphors is a real solid state luminescence. Apart from the slight band displacements, the ZnS:Mn indicates the known orange PL-bands. The emission process of the triboinduced luminescence is bound to the luminescent centres known from the PL. The emission takes place within the Mn^{2+} ion seated on a Zn^{2+} place, without the electrons having left the atom before. It is based on the transition from the ^4G state to the ground state ^6S of the Mn^{2+}. The energy required for the excitation of the Mn^{2+} ion is transferred either directly or indirectly by a resonance process.

3.1.1.2 TL-Spectrum of ZnS:Cu

The luminescence properties of ZnS:Cu, as compared with that of the strongly doped ZnS:Mn, are much more diversified and therefore more complicated. As a rule, there is not only one band occuring in the emission spectrum of the PL, but two bands simultaneously, provided certain preparation conditions are complied with.

As two luminescent centres are in conformity with the two bands, it follows that the Cu atoms are able in the annealing process to build up various centres. In 1961 VAN GOOL found the experimental conditions for the development of the one or other centre. The concentration quotient between the activator atoms Cu and the so-called coactivator atoms of the III — (Al, Ga, In) and VII — (Cl, Br, J) group has a major effect on this formation. In the light of these aspects, the TL, EL and PL-spectra of the three characteristic luminophores ZnS: 10^{-4} Cu, Cl; ZnS: 10^{-3} Cu, J and ZnS: 10^{-3} Cu, Cl were plotted by MEYER, OBRIKAT. The low TL-intensity towards ZnS:Mn required an increase of the grinding quantity from 1 to 150 g. Similar quantities had to be used for the phosphors (Zn, Cd)S: 10^{-4} Cu, Cl; ZnS:Cl and ZnS:Ag, Cl.

Figure 4a shows the spectra of the hexagonal ZnS: 10^{-4} Cu, Cl. TL and PL spectra vary with this substance. Whereas the PL has the green band typical for slight Cu-doping, with the emission peak at 525 nm, the TL-spectrum may be decomposed into two subbands, into a green one with the peak at 520 nm and a blue one at 453 nm. The intensities are 1:0.33. In conformity with the ZnS:Mn here, too, the TL-band is displaced towards the PL-band, however by 5 nm in the direction of lower wavelengths.

Figure 3 shows the experimental results for ZnS: 10^{-3} Cu, J and ZnS: 10^{-3} Cu, Cl. Both luminophores emit a green (525 nm) and a blue band (461 or 470 nm) in the PL. Their intensities are $Q = I_{gr}/I_{bl} = 1:1.41$ and $1:0.12$ respectively after the band decomposition. The TL and PL spectra differ also with these two luminophors. Firstly, both the green (522 nm) and the blue bands (455 or 462 nm) have shifted to the direction of lower wavelengths, and secondly, the TL relative intensities $Q = I_{gr}/I_{bl}$ have shifted for the benefit of the blue bands (ZnS: 10^{-3} Cu, J, e. g. $Q = 1:3.7$). In the region $\lambda = 580$ nm, the course of the intensity of the TL of ZnS: 10^{-3} Cu, Cl indicates an Mn contamination of the sample. Moreover, there was a visible gradation of colours due to the treatment of this luminophor within the jet mill. Whereas the power jet of the mill showed more blue, the mill edge appeared rather green. For the power jet and the mill edge we obtained $Q = 1:1.73$ and $1:1.06$ resp.

For the determination of the origin of this effect, EL-spectra were plotted with variations of the field frequency between 16 c/s and 25 kc/s. For ZnS: 10^{-3} Cu, Cl, these spectra are shown on Figure 3d for 3, 8, 15, and 25 kc/s. We can see a considerably dependence of the intensity quotient Q on the field frequency. In particular it was possible for both TL-spectra (mill edge and power jet) to be brought to coincidence with the EL-spectra of respective field frequency if one disregards the shifting of the TL-bands in the direction of lower wavelengths. The identity concerning $Q = I_{gr}/I_{bl}$ between TL and EL can be taken from Figure 3d for 3 and 15 kc/s. This result demonstrates that the processes taking place in the TL are of electric nature. The equality of the TL and EL-spectra can be explained free from contradictions only if the excitation processes of both kinds of luminescence are identical or at least similar. On this base, the TL of the zinc sulphides must be considered as tribo-induced electroluminescence. The mechanical treatment of the crystallites apparantly results in voltages within the solid which render possible an electric excitation of the luminophor. For an explanation of the different TL-spectra in the power jet and on the mill edge we must proceed from the change of the EL-spectrum as a function of the field frequency. According to BROSER, GUMLICH, MOSER this effect is interpreted by the model of the "delayed recombination". The

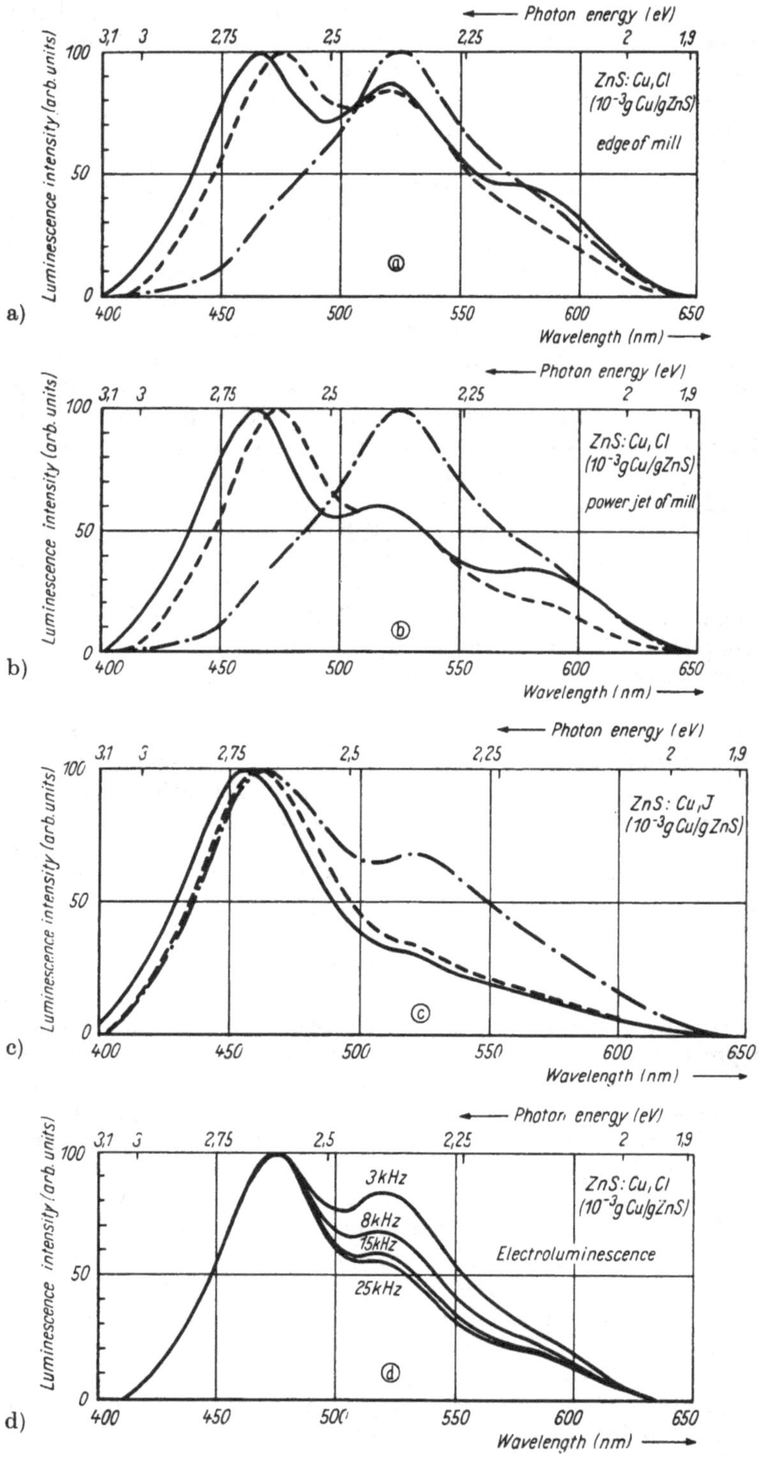

Fig. 3. Comparison of TL–EL- und PL-spectrum of ZnS: 10^{-3} Cu, Cl and ZnS: 10^{-3} Cu, J. (——— triboluminescence, - - - - electroluminescence, – · – · – photoluminescence

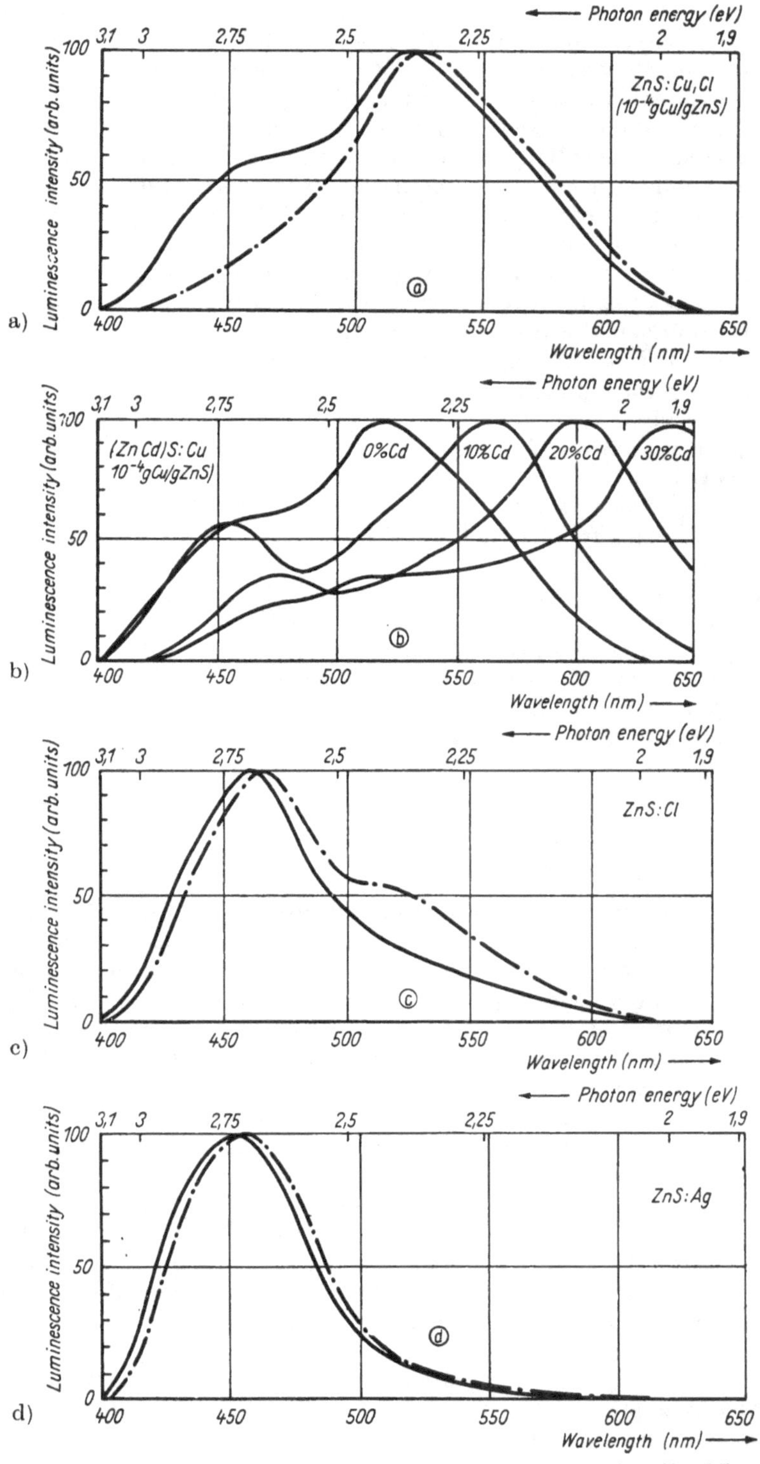

Fig. 4. Comparison of TL- and PL-spectrum of (ZnCd)S: 10^{-4}Cu, Cl; ZnS: Cl and ZnS: Ag. (triboluminescence, – · – · – photoluminescence)

fact that the reaction partners responsible for the luminescence process, i. e. defect electrons and electrons, are separated during half a field period $t = 1/2\ T$ (at sinusoidal excitation between the maximum and minimum of the voltage) is an expression of the delayed recombination. Only at reversal of the field, i. e. during the reanewed reaching of the voltage maximum, they are reunited and recombined spatially, with wavelengths characteristic for the recombination centre being emitted. The time period of the separation of the charge carries can be determined by the selection of the field frequency. This includes at the same time a regulation of the time during which an energetic re-distribution of the defect electrons due to the Schön-Klasens theory is possible. At short time periods (high frequency) there can be no transport and the capture cross sections of the blue centre will control the distribution. In the case of the ZnS:Cu the blue centres have the larger defect electron capture cross section. If the electrons return quickly by field reversal, they will encounter mainly blue centres ready for recombination, in conformity with the experiment. If, however, the separation of the charge carriers is maintained for a sufficiently long time period, the defect electrons will be liberated thermally and a re-arrangement of the holes from the blue to the green centres will take place. With increasing time (decreasing frequency) the fraction of the green bands on the total emission increases.

The relationship shown between the time t of the separation of charge carriers and the change of the El-spectrum of ZnS:Cu may also be transferred to the TL-spectra of this substance. Let us assume the voltage in the mechanical treatment of the zinc sulphide to have been generated by the piezoelectricity, then by analogy with EL, electric fields might occur in both directions, depending on whether there is compression or dilatation of the crystallite. The TL-spectrum then depends

a) on the mechanical pre-history of a treatment of a crystallite contributing to the spectrum, and

b) on the given deformation velocity.

Proceeding from the deformation velocity existing in the jet mill, Kürten and Rumpf stated stress times of $t = 10^{-7}$ s for grains of 40 μm size. Since t in addition to the stress time also is the time of the separation of charge carriers, an equivalent frequency $f = 1/2\,t = 5$ Mc/s follows for the TL spectrum. The TL-spectrum would coincide with the EL-spectrum of this frequency. The intensity quotient between green and blue bands would have to amount to $Q = 1:2.3$. Such a high proportion of the blue bands on the total emission, however, could not be found in the TL. For an explanation of the real Q-values 1:1.73 (power jet) and 1:1.06 (mill edge) we have to assume, therefore, that a second TL radiation component of a relatively stronger green band comes in for the emission at $Q = 1:2.3$. This second component consists of the TL-emission of those ZnS-crystallites which are mechanically treated in the mill two or several times. If, for example, the second treatment takes place after one single rotation of the grain in the jet mill, the time difference between the first and second excitation will be $t = 1 \cdot 10^{-3}$ s. We have assumed a particle velocity $v = 100$ m s^{-1} and a trajectory of 10 cm.

The TL-emission resulting from the second excitation, therefore, is composed of two light pulses. The first light pulse emitted during the increase of the mechanical stresses would coincide with an EL of the frequency $f = 1/2 \cdot 10^3$ s

= 500 c/s and amount to $Q = 1:0.66$. The second light pulse emitted after about 10^{-7} s coincides in colour with an EL of $f = 5$ Mc/s ($Q = 1:2.3$) and fuse together with the first pulse to form a uniform total pulse due to the inertia of the electron processes (decay). If the two pulses show equal radiant intensities, there will a resultant ratio of $Q = 1:1.5$ between the green and blue band. As the total emission in a power jet of the mill is composed of the TL-emission of the first (one pulse at $Q = 1:2.3$) and second excitation (two pulses at $Q = 1:0.66$ and $1:2.3$) we would obtain $Q = 1:1.75$. This value corresponds to an EL-frequency of $f = 15$ kc/s and coincides with the TL-spectrum obtained. This coincidence proves that the TL-spectrum of the power jet can be explained by means of the frequency dependence of the EL-spectrum by assuming a twice repeated treatment of each crystallite contributing to the light emission.

Analogous considerations lead to the interpretation of the TL-spectrum at the mill edge. Contrary to the spectrum discussed above, the proportion of the green band on the total emission here is much greater, being $Q = 1:1.06$. We must assume, therefore, that all crystallites impinging on the mill edge were excited mechanically before. With this assumption, the above mentioned first excitation, which has a small green-blue ratio, does not contribute to the TL-emission on the mill edge. This assumption appears to have been introduced by force and is very plausible because the impact grain-mill edge is possible only by the compressed air jet, i. e. by the power jet of a nozzle and the treated and to be treated ZnS-grains respectively are contained in the power jet (see Fig. 32). Therefore, only grains excited already in the power jet arrive at the crossing-point of the power jet – mill edge, in particular under conditions of larger quantities of material to be milled. Therefore, the TL-spectrum is exclusively composed of the two light pulses of the second excitation. The spectral distribution of the first pulse corresponds to an EL at $f = 1.6$ kc/s with $Q = 1:0.85$, because the time $t = 3 \cdot 10^{-4}$ s passes between the first and the second excitation. We proceeded again from a particle velocity $v = 100\ \mathrm{ms}^{-1}$ and a pathway of 3 cm. Let us assume $Q = 1:2.3$ for the second light pulse, although the stressing time should be longer due to a relatively more tangential movement component toward the mill edge, then follows $Q = 1:1.58$ for the spectrum. The value $Q = 1:1.06$ found experimentally indicates that the radiant intensity of the second pulse can only amount to two third of the first one. This conclusion is the consequence of a small degree of excitation on the mill edge. The treating energy available at the mill edge per grain is oviously smaller than in the power jet. This is expressed by the visibly smaller light intensity and the poorer film density in recording the spectrum.

The TL-spectra of ZnS: 10^{-3} Cu, J and ZnS: 10^{-4} Cu, Cl may be explained in analogy with the discussion of the ZnS: 10^{-3} Cu, Cl. Here, too, the preferred emission of the blue band is a consequence of the high-frequency electric excitation of the TL.

3.1.1.3 TL-Spectra of (ZnCd)S:Cu

The (ZnCd)S-luminophores, which form an uninterrupted solid solution series from pure ZnS to pure CdS, emit broad Pl-bands upon activation with Cu, Ag or upon self-activation. Each of the three bands shifts to ultra-red with increasing Cd-content, as stated by ORTMANN, TREPTOW, because the lattice is expanded and the fundamental lattice absorption moves towards lower energies. MEYER, OBRIKAT investigated the problem if this band shifting also occurs with the TL. Hexagonal (ZnCd)S:10^{-4} Cu, Cl was selected as test substance. As the spectrograph at disposal operates satisfactorily only up to a wavelength of 650 nm, it was only possible to investigate samples of 0, 10, 20, and 30 Mol% Cd. The emission maxima were found to be at 525, 572, 609 or 650 nm. Figure 4b shows the TL emission spectra. We find that even the TL-maxima (520, 567, 602, and 640 nm) shift towards higher wavelengths under conditions of changing Cd content. Figure 5 shows the comparison of the PL and TL emission maxima as a function of the Cd-content. In conformity with the other luminophores here, too, a shift occurs between the TL and PL-bands, amounting to 5, 5, 7 and 10 nm respectively.

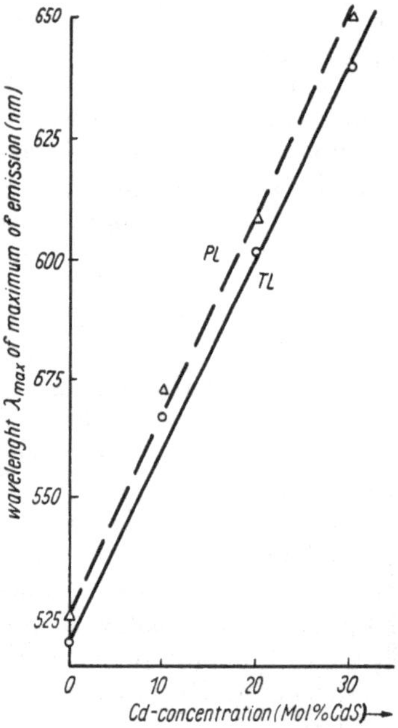

Fig. 5. Dependence of triboemission maxima and photoemission maxima of (ZnCd)S: 10^{-4} Cu, Cl of the Cd concentration

In addition to the (ZnCd)S:Cu bands the authors also obtained the green and blue ZnS:Cu bands (see Fig. 4b) with all substances. This result can only be attributed to a bad preparation, e. g. an insufficient mixture of the two components. Thus, isle-shaped ZnS:Cu centres are formed in the (ZnCd)S. As discussed in 3.1.1.2, these centres will preferably emit the blue band due to the electric excitation. The assumption of an inhomogenous mixture of the initial components ZnS and CdS is supported by the fact that an increasing Cd-content causes the radiant intensity of the ZnS:Cu emission to decrease. The probability for the production of such centres would have to decrease with increasing Cd-addition.

3.1.1.4 TL-Spectrum of ZnS:Cl

The TL-investigation of PL emission band of the zinc sulphide caused by self-activation was conducted on a sample of hexagonal structure. This PL-band situated in the blue region at approximately 465 nm is observed alone only if there are less than 10^{-6} Cu or Ag impurities per g ZnS. The PL-spectrum shown on Figure 4c, therefore, indicates slight Cu-traces, because a green Cu-band (525 nm) occurs in addition to the blue band of the self-activation (466 nm). It is interesting to find that this Cu-band practically cannot be detected in the TL-spectrum on Figure 4c where only a blue band of the emission peak at 462 nm appears. This effect is to be attributed to the preferred TL-excitation of "blue" Cu-centres described in 3.1.1.2. With this luminophor, too, a band shift towards the PL-band of 4 nm in the direction of lower wavelengths is to be detected. The asymmetric shape of the band and its half-width of 0.42 eV are maintained.

According to Era, Shionoya, Washizawa, the emission of the self-activated zinc sulphide is based on the luminescence model of Williams, Prener, i. e. on a pair-occurance of acceptors and donors. They conclude from polarization and electron spin resonance measurements that the centre consists of a Zn^{2+} vacancy and a charge compensating Cl atom seated on a neighbouring lattice site. The luminescence transition takes place between the donor-similar Cl-term and the acceptor-similar Zn^{2+} vacancy.

3.1.1.5 TL-Spectrum of ZnS:Ag

In the incorporation of $10^{-6}-10^{-4}$ g Ag/g ZnS in the zinc sulphide lattice, a blue PL-band (456 nm) is emitted at UV-excitation. The emission process, according to the Riehl-Schön-Klasens model, takes place as an electron transition between conduction band and activator. As the TL-spectrum (Fig. 4d), apart from the shift by 4 nm in the direction of lower wavelengths, is identical with the PL-spectrum, the same emission process may be assumed for the TL.

3.1.1.6 The Shift of the TL-Bands

If the Zn in ZnS is replaced by Cd, the edge of the fundamental lattice absorption shifts towards lower energies, as stated under 3.1.1.3, the lattice is expanded and the Cu and Ag-bands shift into the ultra-red. In the case of the (ZnCd)S:Cu, such a shift was detected also for TL. It is interesting to find that this migration is governed by the same physical causes as the shift of the TL-bands towards the PL-bands found with all spectra. Both phenomena can be attributed to a change of the lattice constants. Whereas the migration of the ZnS:Cu and ZnS:Ag emission bands (at incorporation of Cd) in the direction of higher wevelengths is due to an increase of the lattice distances, the shift of the respective TL-bands towards the PL-bands in the direction of lower wavelengths can be explained by a reduction of the lattice distances. The reduction of the lattice distances results from the strong mechanical treatment of the ZnS and (ZnCd)S respectively where high pressures exist at the moment of the TL-excitation. Under these conditions, the break stress of the ZnS $\sigma = 7.5 \cdot 10^3$ kp cm^{-2} = 7.5 kbar may occur as maximum pressure. In fact, according to the experimental and theoretical studies of Leisaar, Kirs; Ortmann, Treptow and Shionoya et. al., the position of the emission peaks of the ZnS has turned out to be pressure-dependent. The authors mentioned caused hydrostatic pressures up to 52 kbar to act on the luminophores and recorded the PL emission peaks of ZnS:Ag, Al; ZnS:Cu, J; ZnS:Cl, and ZnS:Mn as a function of the pressure. Under these conditions, the energy corresponding with the peak will increase with the Cu and Ag bands, i. e. with the TL the bands migrate towards the lower wavelengths in line with our results.

Owing to a different structure of the centres, the behaviour of the Mn-band differs from the remaining bands. In conformity with the TL-spectrum here, too, we can find a shift in the direction of higher wavelengths. The correlation between band shift and pressure makes it possible for us to state a mean pressure existing in the TL (power jet of the mill) at the time of emission. From the band shifts of the blue Cu (7 nm), green Cu (5 nm), self-activated (4 nm), Ag (4 nm), and Mn-bands (1.5 nm) we obtain pressures at 3,7; 2.8; 1.7; 3.2, and 2.5 kbar which are between half and a quarter of the ultimate stress. The smaller stress value for ZnS:Mn is due to the longer persistence of the TL of ZnS:Mn. The stresses may balance within this time so that even a short time after the fracture and/or the stress maximum the Mn-band emits at normal pressure and the shift is reduced. Let us summarize: The TL-bands show characteristic shifts in relation to the PL-bands. The cause of these shifts may be the high mechanical pressure at the time of the TL-emission.

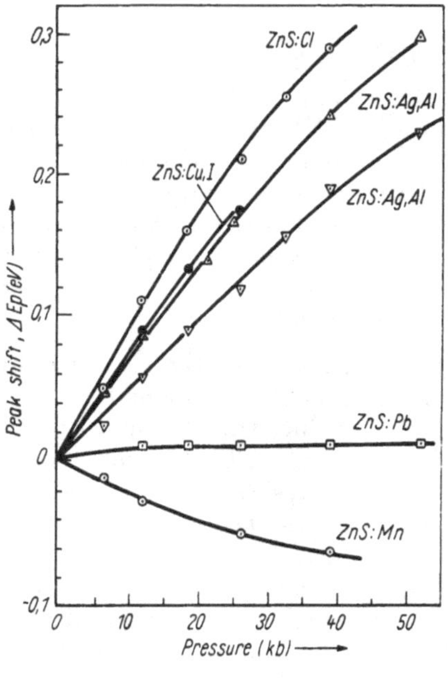

Fig. 6. Influence of pressure on the shift of the maxima of emission (from SHIONOYA et. al.)

3.1.2 Pressure-Light-Correlation

Beside the emission spectrum the correlation between pressure and light is an outstanding feature of the TL of zinc sulphides. From the correlation and the TL radiant intensity as function of the maximum pressure during a mechanical treatment, we can draw essential conclusions for the TL-mechanism on the basis of 3.1.1.2. It will be found that the stress has the same effect on the TL as the radiation intensity on the PL or the voltage or field strength on the EL respectively.

3.1.2.1 ZnS:Mn

The influence of the pressure increase or reduction on the TL is assessed differently in the relevant literature. Whereas ALZETTA et al. (1962) discovered light both in the increase and in the reduction of the stresses, SODOMKA (1964) did not find any TL-emission at the time of pressure increase. In connection with the conclusions drawn from the TL emission spectra concerning the excitation of the TL, and the latest study of ALZETTA et al. (1967) on the brightness waves of the TL and EL at periodic excitation, the results of SODOMKA must be considered as not typically for ZnS:Mn. Therefore we conducted our own measurements on differently doped ZnS:Mn. The compressive stress device shown on figure 7 made it possible for TL and EL investigations on the

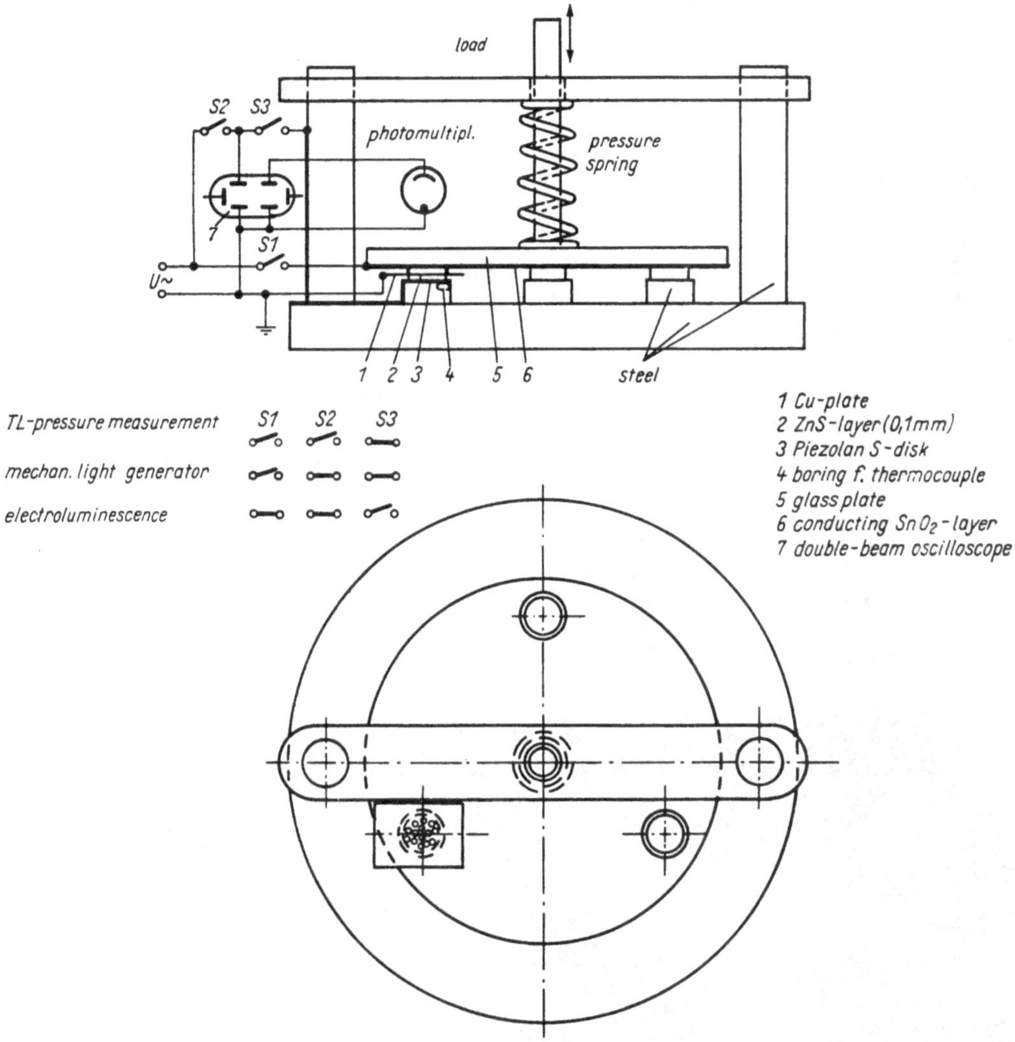

Fig. 7. Apparatus for investigation TL by a load falling down (stress treatment), and by mechanical oscillation produced with piezoelectric ceramics (mechanical light generator) – (The same apparatus allows the study of EL)

same ZnS-sample to be conducted simultaneously or successively. The main component of this device is the circular base plate (110 mm diam.) with the three fitted steel cylinders (10 mm diam.). The front surfaces of the triangularly arranged cylinders form a face-ground surface. A piezoelectric ceramic disk — Piezolan S — is arranged on one of these cylinders for the pressure measurements, moreover a 0.2 mm copper plate serving as electrode, and the luminophor to be investigated. Spacing pieces of electrically insulating material on the two remaining cylinders provide for a plane position of an approx. 5 mm glass plate whose bottom face has a sufficient conductivity at excellent transparency. The treatment of the ZnS-luminophore is effected by dropping a weight. In this procedure, the reproducibility of the TL will be ensured only if the glass plate is prevented from being liftes from the supporting points due to sound wave reflexions, as well as from lateral displacement, the latter effect being encountered by a strong pressure spring or a lateral guide. The light emission was observed from the top by means of a multiplier. In the respective switch position of S1—S3 (compare Fig. 7) it was possible to observe with this device

a) simultaneously the TL and the respective mechanical pressure pulse by dropping a weight,
b) periodic TL (mechanical light generator) or single TL-pulses, depending on the form of voltages applied on the piezo-ceramics, and
c) EL by applying voltages on the copper plate and on the conductive glass surface, without any change of the arrangement.

Figure 8 shows an oscillogram which is typical for the function of the compressive stress device. The figure shows the course of the force developing due to the elastic collision of a steel weight against the glass plate, as a function of time. The time scale amounts to 0.5 ms per abscissa scale division. We can see on the figure how the force leading to the compression of the Piezolan rises steeply, reaches about the maximum after 100 μs, and then decreases again. The steel weight is reflected by the glass surface, leaving the device to itself at the moment of the separation. As the complete device constitutes an oscillatory system, the force maximum is followed by a damped vibration at the frequency $f = 2$ kc/s. This dying-out transient process provides for a second compression of the Piezolan after the dilatation between 0.3 and 0.6 ms, and thus for a repeated treatment of the ZnS luminophore.

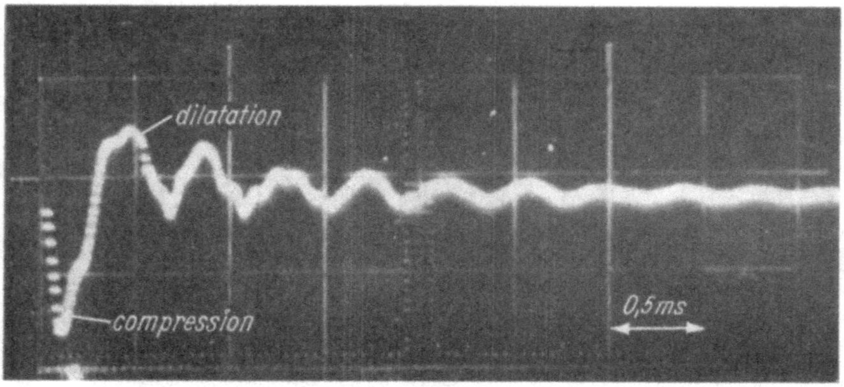

Fig. 8. Shape of pressure in the case of stress treatment

3.1.2.1.1 Comparison of the Light and Stress Pulses

The TL-emission of a ZnS: $5 \cdot 10^{-3}$ Mn powder treated with this device is shown on Figure 9. The light is shown as a positive pulse and that portion of the force curve that corresponds with the compression of the luminophore is

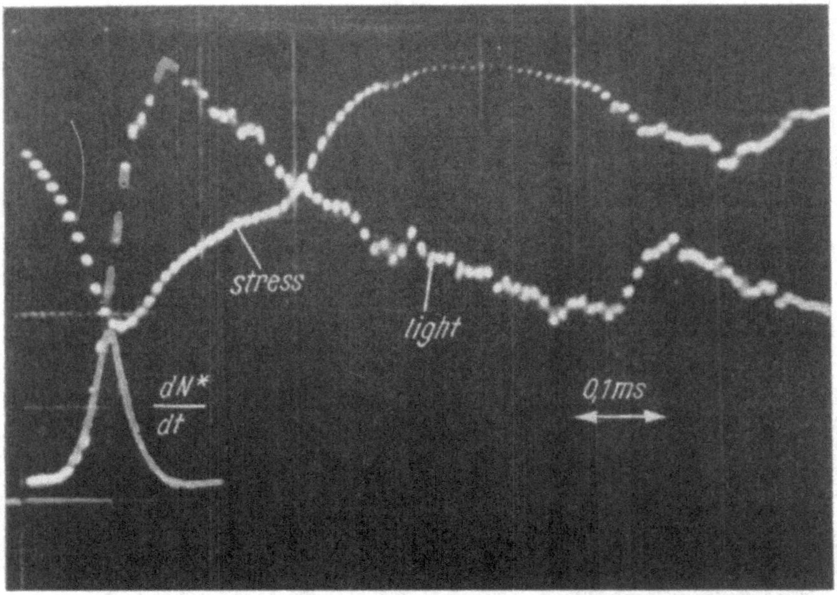

Fig. 9. Comparison of the shape of pressure and TL of ZnS: $5 \cdot 10^{-3}$ Mn

shown as a negative pulse. In conformity with ALZETTA et al. we can see that the TL of the ZnS:Mn occurs both during the increase and during the decrease of the stresses. At the time of the pressure maximum, i. e. 110 µs after the begin of the treatment, the TL-intensity reaches about half the maximum TL-intensity (after 170 µs). On exceeding the maximum, the TL-excitation will stop and the emission will decay exponentially at a time constant of $\tau = 800$ µs (see 3.1.3). The exponential law implies that the number dN of the Mn^{2+} ions excited in the time dt is proportional to the number of the existing excited ions $N(t)$ and the time dt. Then

$$dN(t) = -\frac{dt}{\tau} N(t) . \qquad \text{i. e.} \qquad \frac{dN(t)}{dt} = -\frac{1}{\tau} N(t) .$$

The time decrease of the excited ions $\frac{dN(t)}{dt}$, i. e. the recombination rate, is reflected directly by the TL-intensity and is proportional to the number $N(t)$ of the excited Mn^{2+} ions in the TL of the ZnS:Mn. Since the proportionality between the recombination rate $\frac{dN(t)}{dt}$ and $N(t)$ does not only apply to the TL-decay but forms a universal rule, the rise of the TL-pulse may point to the excitation time t_E and $N(t)$. As follows from Figure 8, the excitation time of 100 µs in the first order approximation is small as compared with the decay time constant $\tau = 800\,\mu s$. From this fact follows

$$-\,dN = \frac{t_E}{\tau}\,N(t_E) \ll N(t_E)$$

and the justification of considering $N(t)$ as the number of excited ions integrated within t_E. Owing to the proportionality

$$I_{TL}(t) \sim -\frac{dN}{dt} \sim N(t)$$

the excitation rate $\frac{dN^*}{dt} > 0$, i. e. the number dN^* of the Mn^{2+} ions excited within the time dt can be obtained by way of simple differentiation of the TL-intensity between 0 and t_E. The function $\frac{dN^*(t)}{dt}$ found in this way has been plotted on Figure 9 in order to better identify the excitation intensity of the TL. In conformity with the EL, most of the Mn^{2+} will be excited when the stress (it is proportional to the voltage with the TL) has reached the maximum. To illustrate this analogy, we have simulated the stress pulse (compression portion) on Figure 10 in the form of an voltage pulse by means of an electric generator

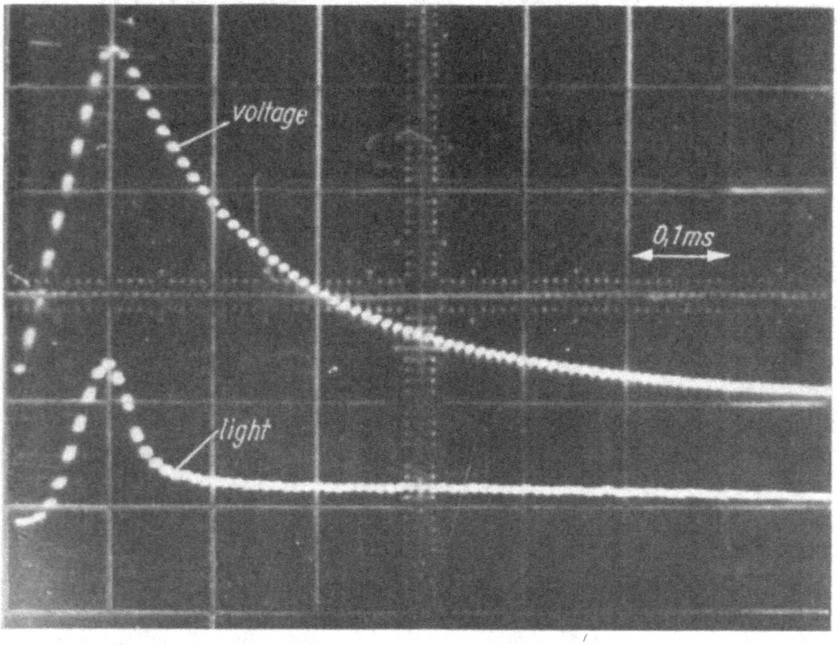

Fig. 10. Voltage pulse and EL-pulse of ZnS: $5 \cdot 10^{-3}$ Mn

and applied to the same ZnS:Mn powder layer. In its course, the EL-emission corresponds with the TL excitation intensity. As the decay of the EL, however, is much faster than that of the TL, there is an agreement of the course of the emission and excitation intensities with the EL. Irrespective of the different TL and EL light pulses, the far-reaching agreement of the excitation intensity proves a close relationship of both kinds of luminescence. This similarity with respect to the excitation mechanism entitles us to consider the TL of the ZnS:Mn as slowly decaying EL.

The symmetry of the excitation intensity visible relative to the stress maximum on Figure 9 makes it possible for us to make a statement on the TL excitation mechanism with ZnS:Mn. The symmetry proves that the TL-emission cannot be explained by the model of the delayed recombination. For an interpretation of the results it will be necessary first of all to assume a direct excitation of the Mn^{2+} ions. This excitation consists of a collision of the "hot" electrons accelerated in the electric field with the Mn^{2+} centres. The emission process following this excitation was described under 3.1.1.1.

As we can gather from Figure 11, the direct excitation may be considered as dominating, but not as the exclusive factor. After the repeated increase of the mechanical stress between 220 and 280 µs, the stress decreased very

Fig. 11. Comparison of the shape of pressure and TL of ZnS: $5 \cdot 10^{-3}$ Mn

rapidly to zero and below that (dilatation). A marked increase of the Tl-intensity is to be observed in this pressure range although the stress — and consequently also the electric field strength — is very small and partly zero. In the range of the initial pressure increase, there is practically no TL-emission under conditions of the same stresses (about 20 µs). This phenomenon can only be explained by the model of the delayed recombination. Beside the direct excitation, energy is also transferred to the Mn^{2+} ions by resonance, which are released during the transition of electrons between imperfections and bands.

In this connection we should like to comment a phenomenon mentioned for the first by CHUDÁČEK (1967). According to his theory, the main light pulse is followed by a second TL-pulse of smaller amplitude after 2.4 ms, following the single treatment with a steel needle. Without tracing the process of treatment by measuring the stresses, the author attributes the first pulse to elastic-plastic deformations and the second one to the fracture of crystallites. A repeated treatment of the luminophore due to sound wave reflexions is not discussed as a second cause of the TL-pulse.

In conformity with the measurement result stated by CHUDÁČEK (1967), our own measurements also showed a second TL-pulse. Such a pulse, for example, appears after 0.65 ms on Figure 9. From the course of the force recorded simultaneously we can see very clearly that this TL-emission is bound to the renewed increase of the mechanical pressure and thus to a second treatment of the luminophore. Upon variation of the natural frequency of the treatment device, the time difference between the first and the second pulse may be freely selected within wide limits. On Figure 12 the difference amounts to 1.6 ms. After another 0.35 ms a third TL-excitation takes place on this oscillogram. The intensity quotient between the first and second pulse is generally smaller than 1, because energy is getting lost due to dampening. The concrete value, however, depends on the setup of the treatment device and its material properties.

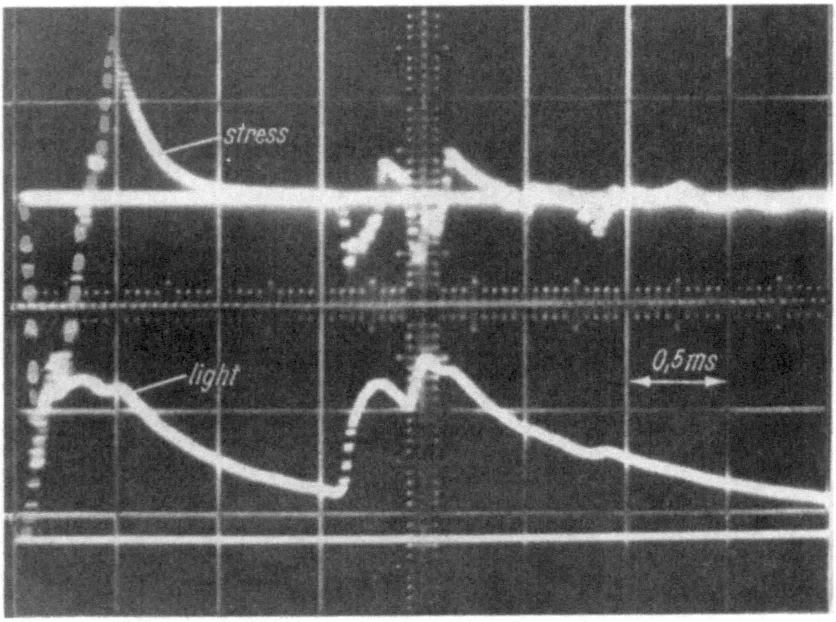

Fig. 12. TL-Pulse of ZnS: $5 \cdot 10^{-3}$ Mn after a second mechanical treatment

3.1.2.1.2 TL-Intensity as a Function of the Stress

Whereas the emission process could be derived from the emission spectrum and the excitation mechanism of the TL from the comparison of the light pulse with the stress pulse, the determination of the function TL-intensity in dependence on the stress offers the possibility of providing a statement on the significance of the stress for the TL. For this purpose, this function as determined by means of the treatment device described and shown on Figure 13. The TL radiation intensity I was plotted as a function of the reciprocal square root of the force K measured and as a function of $\bar{\sigma}^{-1/2}$ respectively ($\bar{\sigma} = K/F$ — average stress). We obtain a direct proportionality between I and $K^{-1/2}$.

The TL-intensity as a function of the force K thus can be expressed as follows

$$I = i_0 \, e^{-b K^{-\frac{1}{2}}} \tag{1}$$

Here, $b = 38.1 \, V^{1/2}$ forms the constant of rise and $i_0 = 2.04 \cdot 10^5$ units of light the maximum possible TL-intensity for ZnS:Mn. Voltage were applied to the same ZnS-sample and the voltage dependence of the EL was measured. The EL-intensity I is shown on Figure 14 as a function of $U^{-1/2}$. At an EL cell thickness of 10^{-2} cm the voltage variation between 125 and 400 V corresponds to a range of the electric field strength of 1.25 to $4 \cdot 10^4 \, V \, cm^{-1}$. In conformity with many authors, there is a linear relationship between I and $U^{-1/2}$ within this range:

$$I = I_0 \, e^{-B U^{-\frac{1}{2}}} . \tag{2}$$

The constant of rise here is $B = 184 \, V^{1/2}$ and the maximum EL intensity $I_0 = 1.4 \cdot 10^7$ units of light. As the above discussion stated the TL to be triboinduced EL, the two equations (1) and (2) may exist side-by-side under these conditions and only under these conditions, provided K and U are proportional to each other

$$K = \alpha \, U . \tag{3}$$

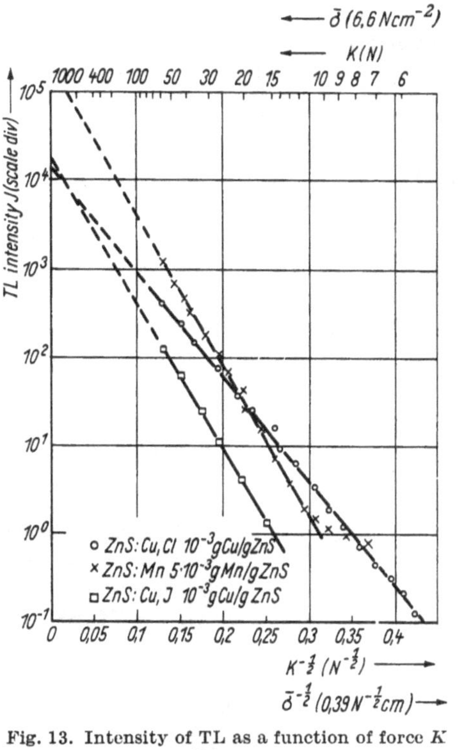

Fig. 13. Intensity of TL as a function of force K and the mean stress $\bar{\sigma} = K/F$

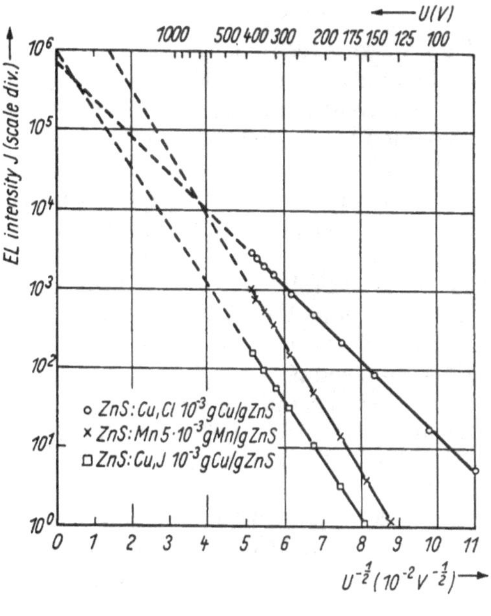

Fig. 14. Intensity of EL as a function of voltage

The proportionality constant α can be determined from the constants of rise b and B.

The following equation applies

$$-\frac{b}{\sqrt{K}} = -\frac{B}{\sqrt{U}} \tag{4}$$

on the assumption that the tribo-induced voltages $U = 1/\alpha\, K$ act on the ZnS-luminophore in the same way as the externally applied voltage. From the formulation (4) follows the value for

$$\alpha = \frac{b^2}{B^2}. \tag{5}$$

3.1.2.2 ZnS:Cu

Beside the dependence of the emission spectrum on the field frequency (3.1.1.2) another consequence of the delayed recombination is expressed by the fact that no light can be noticed upon initially applying an electric field to an electroluminescent ZnS:Cu phosphor, because a voltage rise causes only defect electrons and electrons to be separated from each other. The high mobility of the electrons prevents an immediate recombination in the excitation region. Light emission is possible only at field reversal. Contrary to this, a strong EL and TL emission is observed even with an initial voltage rise under conditions of a direct excitation by electron collision (ZnS:Mn). We can see that electroluminescent ZnS:Cu and ZnS:Mn may have completely different EL and TL properties due to different excitation mechanisms.

3.1.2.2.1 Comparison of Light and Stress Pulses

The above aspect of the delayed recombination has an extremely important and experimentally verifiable consequence for the TL. In the first line, a pulse-shaped treatment can only cause light to be emitted when the stress has exceeded the maximum, i. e. $d\sigma/dt < 0$.

The result characteristic for electroluminescent ZnS: 10^{-3} Cu, Cl can be seen on Figure 15 in the form of an oscillogram. The figure shows the course of the light and that portion that corresponds to the compression of the ZnS as a positive pulse. We clearly identify two light intensity maxima I_1 and I_2, with $I_2 \gg I_1$. Purely phenomenologically we can assign I_1 to the compression maximum and I_2 to the dilatation maximum. The intensity difference between I_1 and I_2 justifies the statement that the TL of the electroluminescent ZnS: 10^{-3} Cu, Cl is observed preferably with a dilatation $d\sigma/dt < 0$ following the compression, if the ZnS-crystallites were not excited mechanically or electrically for a longer period before this treatment. In order to make the analogy between TL and EL more distinguishable, the stress pulse was simulated analogous to the ZnS:Mn by a voltage pulse and applied to the luminophorous layer. The EL-light pulse on Figure 16 shows a far-reaching agreement with the TL-pulse.

On the basis of this agreement, the processes taking place in the mechanical impact treatment of ZnS: 10^{-3} Cu, Cl may be interpreted as follows: Upon increase of the stresses and thus of the voltages, different processes are taking place in the regions of the types A and B of a crystallite. Whereas in the regions of the type A, the excitation process of the triboinduced EL takes place

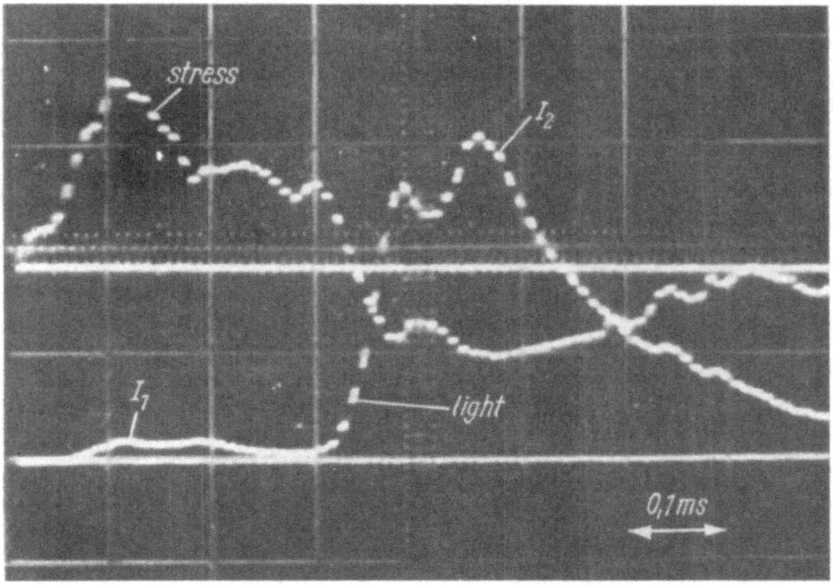

Fig. 15. Comparison of the shape of pressure and TL of ZnS: 10^{-3} Cu, Cl

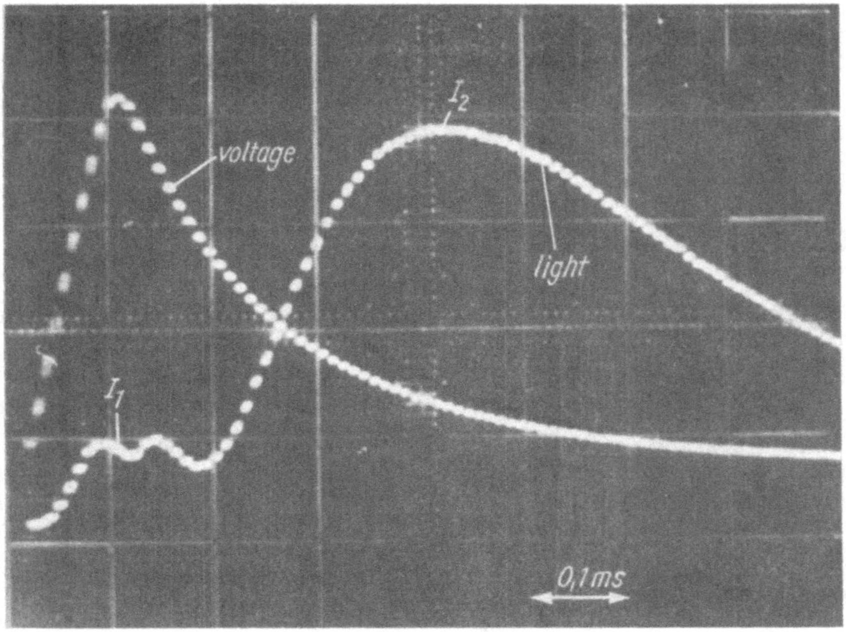

Fig. 16. Voltage pulse and EL-pulse of ZnS: 10^{-3} Cu, Cl

in the form of the separation of electrons and defect electrons, with the electrons migrating to the anode and dropping into traps there, the electrons and defect electrons situated at the points of the type B (localized in the activator atoms) are recombined, with I_1 being emitted. As, however, contrary to the classical

EL this TL-pulse I_1 never decays completely even in a luminophore that was not excited previously for a long time, the amplitude of I_1 can be satisfactorily explained by a partly immediate recombination of the electrons and defect electrons generated. If the stress exceeds the peak value at the moment t_1, if it then becomes zero or negative, the polarity of the tribo-induced voltage will change. The electrons liberated from the traps will become accelerated to the ionized activators ready for recombination in the regions of the type A and generate a strong light pulse I_2 during the recombination process.

Since there are regions in the ZnS-crystal which, as indicated already, emit light only at the moment of a positive or negative polarity and as the excitation process (type A) of the triboinduced EL for the TL-pulse I_2 takes place at the time of the stress increase ($t < t_1$), electrons and defect electrons are separated from each other at the moment of the main emission of the TL ($t > t_1$) in the positions of the type B. If a time Δt passes till the second treatment, the

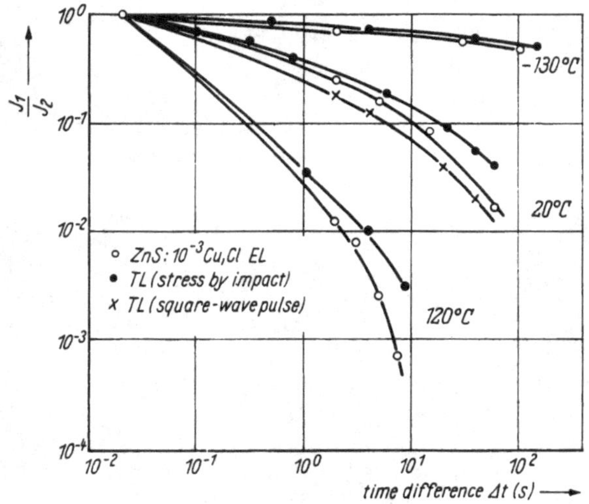

Fig. 17. Magnitude of the first pulse I_1 (related on I_2 = const.) as a function of pulse sequence time Δt of ZnS: 10^{-3} Cu, Cl

radiationless electron transitions taking place within Δt will lead to a reduction of occupied traps and ionized activators. The number of the theoretically possible radiative recombination transitions, which would develop at repeated treatment, are decreasing with increasing Δt. By the selection of a small Δt (a few ms) it would be possible for us to generate a light pulse $I_1 \approx I_2$ during the second treatment already in the range of the pressure increase at the time t_1. The amplitude of I_1 and the intensity quotient I_1/I_2 respectively would have to depend on the time Δt and the temperature because the transition probability of radiationless recombinations is also largely temperature-dependent. As a matter fact, a marked influence of the quotient I_1/I_2 could be found from the mechanical pre-history or more precisely the waiting time Δt between two treatments and the temperature. This result is shown on the diagram of Figure 17. For the temperatures — 130, 30, and 120 °C, the intensity quotient I_1/I_2 of the TL and EL is shown as a function of the waiting time Δt. We obtain a distinct temperature and Δt-dependence as well as a relatively good agreement between the EL and the TL generated by the impact treatment. With all temperatures, however, the TL I_1/I_2-quotients are greater without exception than the corresponding values of the EL. The differences increase considerably with increasing Δt. The cause of this fact, as discussed already, can be explained only by an immediate recombination of the excited electrons with the defect electrons in the excitation region. A totally improved conformity, in particular for the greater Δt, between EL and TL is to be obtained with rectangular TL stress pulses. A Piezolan disk applied to the rectangular 400 V voltage pulses served as force generator.

Beside the strongly electro-luminescent ZnS: 10^{-3} Cu,Cl, pressure light investigations were also conducted on the weaker electroluminescent ZnS: 10^{-3} Cu, J and on the very weakly electro-luminescent ZnS: 10^{-4} Cu, Cl at voltages up to 400 V. The light and stress pulses of these two phosphores are shown in the Figs. 18 and 19 for high Δt and room temperature. As compared with ZnS: 10^{-3} Cu, Cl, ZnS: 10^{-3} Cu, J shows a greater I_1/I_2 quotient, and ZnS: 10^{-4} Cu, Cl a still greater one. Whereas the TL-behaviour of the first two substances is still very similar, because a distinct temperature and Δt-dependence on I_1/I_2 occurs also with ZnS: 10^{-3} Cu, J, the behaviour of the latter phosphor deviates considerably from that of the first ones. For the quotient I_1/I_2 independent of Δt, applies $I_1/I_2 \gg 1$. A marked decrease is to be expected only at higher temperatures. It is interesting to find that the I_1/I_2-increase programming from ZnS: 10^{-3} Cu, Cl, via ZnS: 10^{-3} Cu, J, and ZnS: 10^{-4} Cu, Cl is accompanied by an increase of the PL decay time constant (see 3.1.3) and a decrease of the EL-emissivity. As a matter of fact, this distinct correlation is not accidental, because the increase of the PL decay time constant is based on an increase of the trap depth and because phosphores with energetically deep-lying traps as a rule have a poor EL-emissivity. If excited electrons get into deep traps, they will not be able to leave them again at field reversal due to the relatively great energy required for liberation. A great portion of the excited electrons is bound in traps, thus getting lost for the EL-emission. On the other hand, the presence of deep traps for the same reason partly or completely prevents the effectiveness of the delayed recombination. The delayed recombination cannot take place if a recombination between electrons and defect electrons in the excitation region is made impossible due to the capture of the excited electrons accelerated to the anode. The effectiveness of the delayed recombination, therefore, is bound to decrease considerably at an increasing trap depth and PL decay time constant respectively, with the I_1/I_2-quotient increasing. In a limiting case under conditions of very deep traps, a TL-emission is obtained only at the moment of the TL-excitation by an immediate recombination in the excitation region. As I_2 therefore is very small or even zero, $I_1/I_2 \gg 1$ or ∞. This condition occurs with substance ZnS: 10^{-4} Cu, Cl which shows an extremely prolonged PL-phosphorescence.

3.1.2.2.2 TL-Intensity as a Function of the Stress

Beside the TL-intensity of ZnS:Mn, Figure 13 also shows the TL-intensity of ZnS: 10^{-3} Cu, Cl and ZnS: 10^{-3} Cu, J as a function of the reciprocal square root of the force K measured at the time of the compression maximum. Both Cu-doped zinc sulphides, too, show a strictly linear relation between I und $K^{-1/2}$ and $\bar{\sigma}^{-1/2}$ respectively. In conformity with the ZnS:Mn here, too, the TL-intensity may be expressed as a function of the force by

$$I = i_0 \, e^{-b\, K^{-\frac{1}{2}}} . \qquad (2)$$

The rise constant b and the maximum possible TL-intensity i_0 is 27.4 $V^{1/2}$ and $1.5 \cdot 10^4$ light units for ZnS:10^{-3} Cu, Cl and 37.8 $V^{1/2}$ and $1.7 \cdot 10^4$ light units for ZnS:10^{-3} Cu, J respectively. The voltage dependence of the EL measured on the same samples is shown on the diagram of Figure 14 as a function of $U^{-1/2}$. The correlation between I and U stated in the equation (2) applies to both luminophores. For the ZnS:Cu coactivated with Cl we obtains $B = 110\ V^{1/2}$ and $I_0 = 7.6 \cdot 10^5$ light units and coactivated with J $B = 170\ V^{1/2}$ and $I_0 = 1 \cdot 10^6$ light units.

In complete analogy with ZnS:Mn, equations (1) and (2) may exist simultaneously even under conditions of Cu-doped phosphors, provided K and U are proportional to each other, i. e. there exists the relationship

$$K = \alpha\, U \qquad \text{with} \qquad \alpha = \frac{b^2}{B^2}$$

between both of them.

Owing to an insufficient EL-intensity, the comparison TL—EL could not be conducted for ZnS:10^{-4} Cu, Cl.

3.1.2.3 ZnS:Ag and ZnS:Cl

The correlation between the stress pulse and the TL light pulse of ZnS:Ag and ZnS:Cl is shown on Figs. 20 and 21. It can be seen from the oscillogram that the TL-emission consists only of a light pulse I_1 that reaches its maximum at the moment of the maximum compression. Due to the similarity of the light pulses shown on Figs. 18, 20 and 21, the excitation and emission process of these luminophores must agree with that of the ZnS:10^{-4} Cu, Cl. The luminescent processes discussed under 3.1.2.2 for ZnS:10^{-4} Cu, Cl may also be extended to ZnS:Ag and ZnS:Cl, because all the three luminophores show a prolonged PL-phosphorescence (see 3.1.3) and therefore similarly deep traps and a small EL-emissivity. The TL-emission of these luminophores, therefore, is not affected by the delayed recombination. It develops due to the immediate recombination of the excited electrons in the excitation region. This explains the independence of the light pulse I_1 on a preceding TL-excitation.

Concerning the investigations conducted by Obrikat with respect to the correlation between TL and the stress acting in the ZnS, we can state in summary that the stress — light correlation forms an extremely sensitive instrument for the characterization of the TL-properties of a luminophore of the ZnS-type. In dependence on the temperature, the traps existing in the phosphor, as well its mechanical, electrical pre-history (trap filling), TL-emission occurs either at increase and decrease or only at decrease after preceding increase of the stresses. Never, however, a sole emission was observed at the moment of the stress increase during a compressive treatment. Such a phenomenon would be inconceivable on the basis of the developed model conception within the system of the tribo-induced electroluminescence.

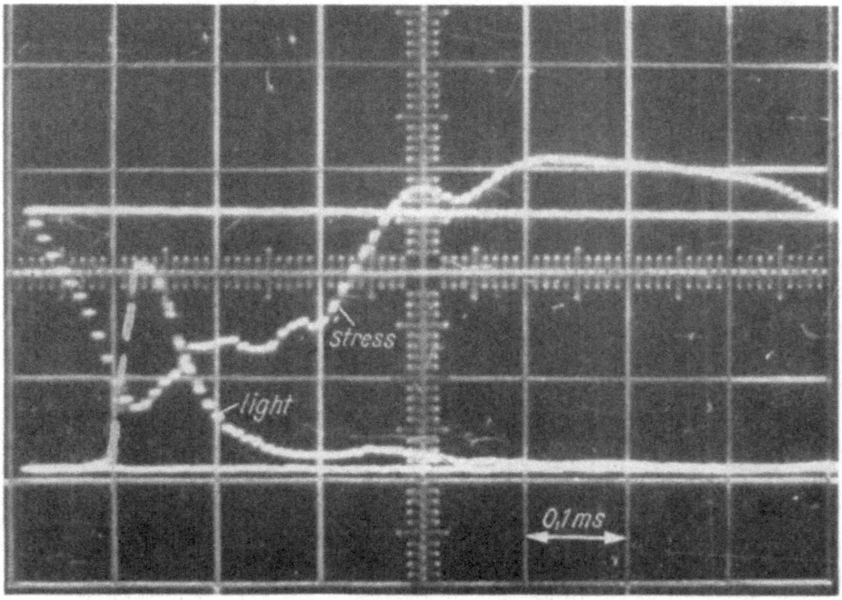

Fig. 18. Pulse of pressure and TL of ZnS: 10^{-4} Cu, Cl

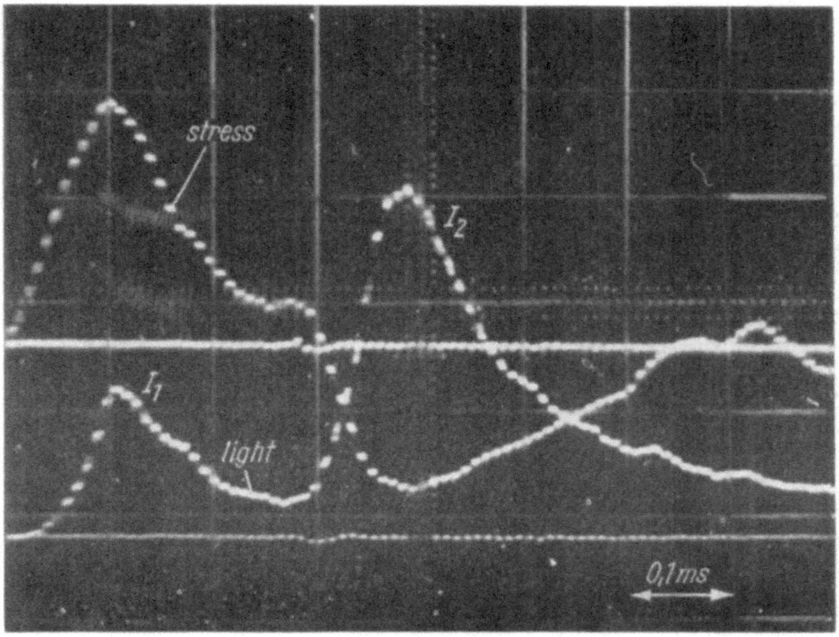

Fig. 19. Pulse of pressure and TL of ZnS: 10^{-3} Cu, J

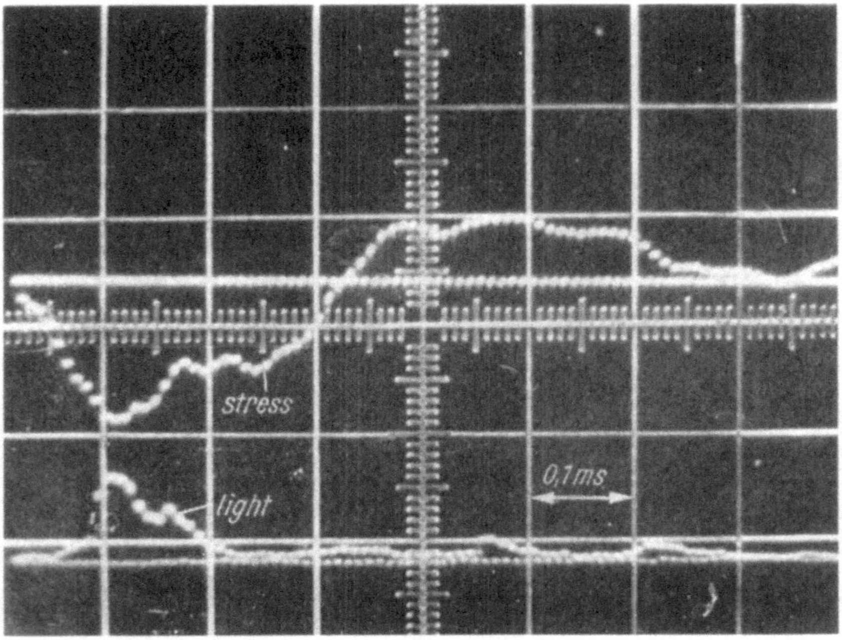

Fig. 20. Pulse of pressure and TL of ZnS: Ag

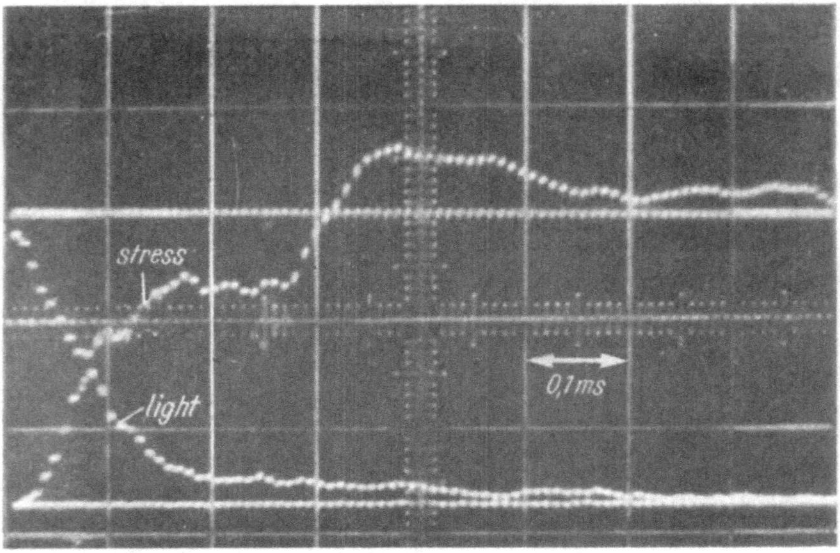

Fig. 21. Pulse of pressure and TL of ZnS: Cl

3.1.3 TL-Decay

After having discussed important TL-characteristics in connection with the emission spectra and the stress — light — correlation in the preceding chapters, we shall now deal with a classical characteristic of each luminescence, with the decay of the light emission after the luminescent excitation. The first TL decay investigations were conducted by Chudáček (1967) on ZnS:Mn, ZnS:Cu, and ZnS:Ag. Similarly to Waggoner and Nelson in the spectrum recording, the author used a modified form of the PL decay circular disk introduced by Wood, and an impact treatment for ZnS:Ag and ZnS:Cu, using a steel pin as a tool.

3.1.3.1 ZnS:Mn

The decay curve of ZnS:Mn shows an exponential course with a time constant $\tau = 0.36$ ms. In the range between 0.5 and 2.5 ms after termination of the excitation, the exponential curve passes over into a flat stage (Fig. 22). Its

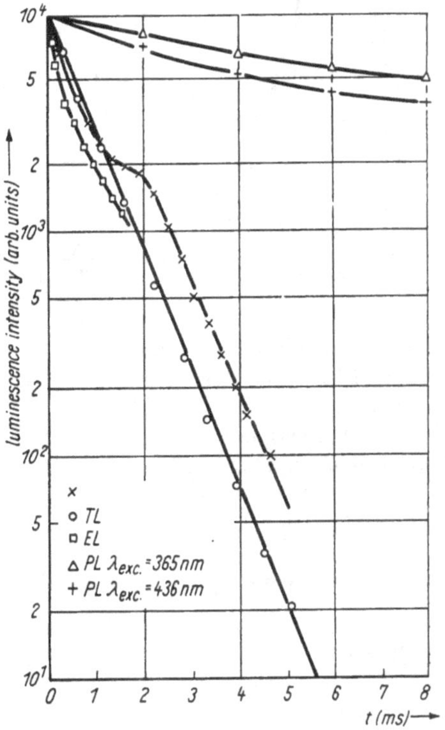

Fig. 22. Decay curves of the TL, EL and PL of ZnS: $5 \cdot 10^{-3}$ Mn

appearence is explained by the second TL-pulse stated under 3.1.2.1, applying an impact treatment. Our own investigations with the decay disk on five ZnS:Mn luminophores doped between $5 \cdot 10^{-4}$ g and $5 \cdot 10^{-2}$ g Mn/gZnS showed that this stage occurs at one time and stays away at another time on the same material and apparantly under the same conditions. It was not possible to find the conditions for the occurence of this stage. Therefore, both TL decay curves, which have the same time constant $\tau = 0.7$ ms, were plotted for ZnS:$5 \cdot 10^{-3}$ Mn on Figure 22. For comparison we have also plotted the EL and PL decay at an excitation wavelength of the irradiated light of $\lambda = 365$ and 436 nm. The time constant of the EL, being 0.37 ms, comes much closer to the time constant of TL than that of the PL ($\tau = 18$ ms). The TL and PL decay curves at different Mn-contents are shown on Figure 23. The time constants of the TL and PL

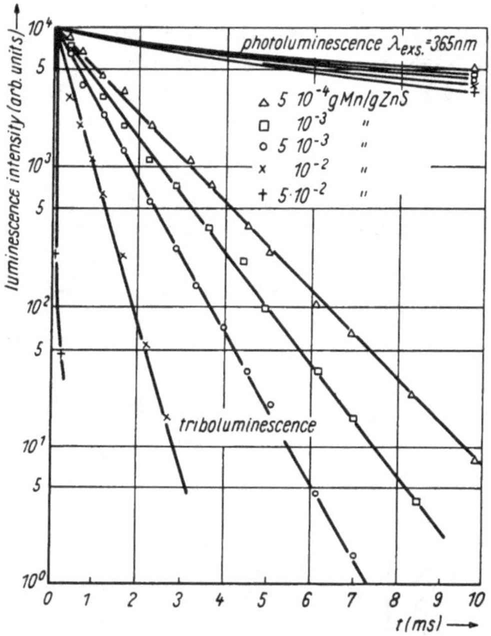

Fig. 23. TL and PL curves of decay of different doped ZnS:Mn powders

decrease with increasing Mn concentration. This effect is pronounced particularly strong at the TL as with $\tau = 1.3$ ms ($5 \cdot 10^{-4}$ Mn) and $\tau = 0.025$ ms ($5 \cdot 10^{-2}$ Mn) the time constant varies like 60:1. With higher Mn-contents ($\leqq 5 \cdot 10^{-2}$ Mn) the exponential character of the ZnS:Mn decay is pushed back and a hyperbolic law is gaining ground. The farreaching validity of the exponential law is in conformity with the emission process outlined under 3.1.1.1, which contrary to ZnS:Cu, ZnS:Ag and ZnS:Cl may be considered as monomolecular reaction.

3.1.3.2 ZnS:Cu, ZnS:Ag, and ZnS:Cl

The problem of decay measurements on phosphors, which have a relatively small time constant τ, is indicated on these ZnS-luminophores. An exact measurement of τ will only be possible if the excitation intensity drops to zero within a time $t \ll \tau$. This condition $t \ll \tau$ probably was not complied with in the measurement on ZnS:Cu and ZnS:Ag conducted by CHUDÁČEK (1967), because the time constants $\tau = 1$ ms and 5 ms stated by him appear much greater. We are in conformity with KÜRTEN and RUMPF that the true TL decay constant cannot be carried out by means of a compressive stress in which the relatively great masses are in the way of a high treatment speed. Exact values can only be obtained by means of a bouncing treatment by shooting the ZnS grain against a baffle plate or a small steel ball (1 mm diam.) at high velocity against the luminophore, and by means of the decay circular disk stated above. Consequently, the τ-values found by KÜRTEN, RUMPF and OBRIKAT are in the orders below 1 ms.

The decay curves shown on Figure 24 were plotted on ZnS: 10^{-3} Cu, Cl; ZnS: 10^{-3} Cu, J; ZnS:Ag, Cl and ZnS:Cl by means of the decay disk. The phosphors in question are without exception those of whom the TL-spectrum and the stress — light — behaviour are known. The electroluminescent substances

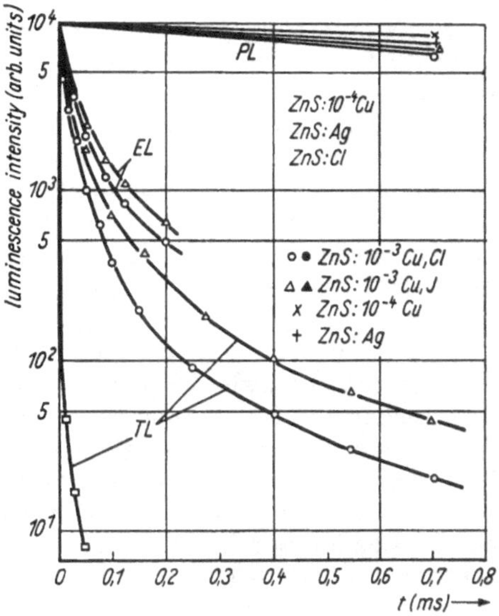

Fig. 24. Decay curves of TL, EL and Pl of ZnS: Cu, ZnS:Ag and ZnS:Cl

ZnS: 10^{-3} Cu, Cl and ZnS: 10^{-3} Cu, J show $\tau = 17$ and 17 μs with the TL $\tau = 26$ and 33 μs with the EL, and $\tau = 6.5$ and 10 ms with the PL resepctively In analogy to the ZnS:Mn, the electroluminescent ZnS:Cu phosphors also result in a resonable agreement between TL and EL. Contrary to the ZnS:Mn the EL decay curves here are above the TL-curves.

The non-electroluminescent samples ZnS: 10^{-4} Cu, Cl; ZnS:Ag, and ZnS:Cl show an equal behaviour of TL as in the case of the stress-light-correlation. The TL time constant is < 2 μs for all τ, whereas τ amounts to 70,15 and 15 ms with the PL respectively. The exact determination of the τ of the TL could not be made because the limit of the time resolution fixed by the measuring setup had been reached at 2 μs. On a ZnS: Cu-sample which probably had the same properties as the ZnS: 10^{-4} Cu, Cl investigated by us, Kürten, Rumpf determined a luminescence period of 10^{-7} s by means of the crushing method. This time is about identical with the time required by the fracture to migrate through a ZnS-crystallite of about 50 μm size at a propagation velocity of approx. 10^3 ms^{-1}. The contact times occuring in the elastic collision of the crystallites and calculable on the basis of the Hertzian collision theory are of the same order or even below. From the agreement detected as far as in the 10^{-7} s—range between the TL decay time constants and the time within which the ZnS-crystallites are deformed follows that in the case of the non-electroluminescent substance group the recombination processes leading to the TL-emission take place within an extremely short time. Contrary to ZnS: 10^{-3} Cu, Cl and ZnS: 10^{-3} Cu, J, the decay here does not occur due to the time required for the electron transport or recombination processes but by the time required by the elastic waves to pass through the crystal. As long as they deform the crystal, a TL excitation will be caused. The decrease of the TL-intensity is not an actual decay but a vanishing of the light emission for the elimination of the cause of the excitation. Such a light emission can be considered as a classical example for a tribo-induced fluorescence.

The picture just outlined for ZnS:10^{-4} Cu, Cl; ZnS:Ag and ZnS:Cl from the TL-decay is in complete conformity with the recombination process discussed under 3.1.2.2 for ZnS:10^{-4} Cu, Cl. The small period of time $\leqq 10^{-7}$ s following from the decay is identical with the time within which the activators be ionized and the defect electrons located there recombine with the excited electrons. The time τ must be as small as that because the stress — light — correlation required an immediate recombination of the excited electrons with the defect electrons in the excitation region (place at which the electrons get into the conduction band), i.e. without a noticeable electron transport. The electrons accelerated considerably due to the high electric field strength thus must be able within a short time to impinge on an ionized activator atom and to recombine at TL-emission because they cover only an extremely short distance.

As against that, the TL-decay of the ZnS:10^{-3} Cu, Cl and ZnS:10^{-3} Cu, J at sufficiently quick interruption of the TL excitation constitutes a property of the electron processes taking place in the phosphor. The delayed recombination forms the basis of these processes. After the electron excitation the major portion of the electrons is moved away into more remote regions, whereas by far the smaller portion recombines radiatively by emitting the TL-pulse I_1 mentioned under 3.1.2.2. The removed electrons, which have dropped into traps, can return to the excitation region only under conditions of an elimination of the electric field and recombine there with the activators ready for recombination. The TL-decay then is conditioned by the time dependence of this recombination rate, i.e. by the finite times required for the liberation of the electrons from the traps and by the electron transport.

Beside these short-time decay measurements, an extremely intensive treatment of approx. 30 g ZnS:10^{-4} Cu, Cl caused such a high TL-intensity to be generated that the TL-decay could be observed over minutes up to a maximum of half an hour. In the period of recording between 1 min and 30 min it was identical with the known PL long-time decay of the green Cu band. The effect can be explained on the basis of 3.1.2.2 as follows. The intensive mechanical treatment in the beater mill has resulted in a high concentration of occupied traps. The concentration is high enough so that the thermal liberation per time unit brings enough electrons into the conduction band and then into ionized activators in order to generate a detectable light intensity. The developing phosphorescence must be identical with the PL-decay.

Detailed experimental data on the TL and PL decay and on the TL, EL, and PL intensities of the luminophores investigated are compiled on table 2.

3.1.4 Temperature Dependence of the TL

The temperature dependence of the TL of ZnS:$5\cdot10^{-3}$ Mn, ZnS:10^{-3} Cu, Cl; ZnS:10^{-3} Cu, J, and ZnS:Ag was recorded in the temperature range of 90 to 500 °K by means of a compression treatment device which guarantees an excellent reproducibility of the TL.

Table 2

Comparison of characteristic TL, EL and PL measurement results

phosphor	hexag. portion	triboluminescence						EL	photoluminescence		
		compressive stress device			jet mill	decay disk					
	[%]	Q	Φ	Q/Φ	Q	$\tau_{e^{-1}}$	$\tau_{10^{-1}}$	$\Phi_{corr.}$	$\Phi_{corr.}$	$\tau_{e^{-1}}$	$\tau_{10^{-1}}$
ZnS:Cu, Cl 10^{-3} g Cu/g ZnS	24	0.08	$4 \cdot 10^2$	200	0.013	17	50	$3 \cdot 10^3$	0.4	6500	30000
ZnS:Cu, I 10^{-3} g Cu/g ZnS	83	0.037	$1.5 \cdot 10^2$	250	0.006	17	50	$1.8 \cdot 10^2$	0.4	10000	50000
ZnS:Cu, Cl 10^{-4} g Cu/g ZnS	100	0.035	$1.5 \cdot 10^2$	230	0.006	<2	<7	$1.1 \cdot 10^1$	2.3	70000	360000
ZnS:Ag	100	0.023	$1 \cdot 10^2$	230	0.005	<2	<7	$1 \cdot 10^1$	2.3	15000	120000
ZnS:Cl	100	0.05	$2 \cdot 10^2$	250	0.012	<2	<7	$1 \cdot 10^1$	2	—	—
ZnS:Mn $5 \cdot 10^{-2}$ g Mn/g ZnS	80	0.096	$6 \cdot 10^2$	160	0.66	25	65	$5 \cdot 10^1$	0.2	10000	50000
10^{-2} g Mn/g ZnS	30	1.8	$3 \cdot 10^3$	600	3.3	320	960	$8 \cdot 10^2$	1.6	17000	57000
$5 \cdot 10^{-3}$ g Mn/g ZnS	10	3.3	$5 \cdot 10^3$	700	3.3	700	1900	$4 \cdot 10^3$	2.6	18000	65000
10^{-3} g Mn/g ZnS	6	1.10	$1 \cdot 10^3$	1100	1.8	1000	2300	$4 \cdot 10^2$	1.6	20000	160000
$5 \cdot 10^{-4}$ g Mn/g ZnS	2.5	0.84	$6 \cdot 10^2$	1400	0.6	1300	3100	$3.5 \cdot 10^2$	1.6	23000	200000

Q = quantity of light (prop. to intensity)
Q/Φ [10^{-6} s]
Φ = light flux
τ [10^{-6} s] = time constant
$\tau_{e^{-1}}$; $\tau_{10^{-1}}$ = time of decay to a magnitude of e^{-1} and 10^{-1}

Figure 25 shows the temperature dependence of the TL of ZnS:$5 \cdot 10^{-3}$ Mn suspended in Lüvakol oil, silicon oil and without dielectric, as well as the EL in silicon oil. The similarity between the EL and TL course without dielectric is apparent. The deviation at low temperatures is probably due to a change of the dielectric properties of the silicon oil. EL measurements could not be con-

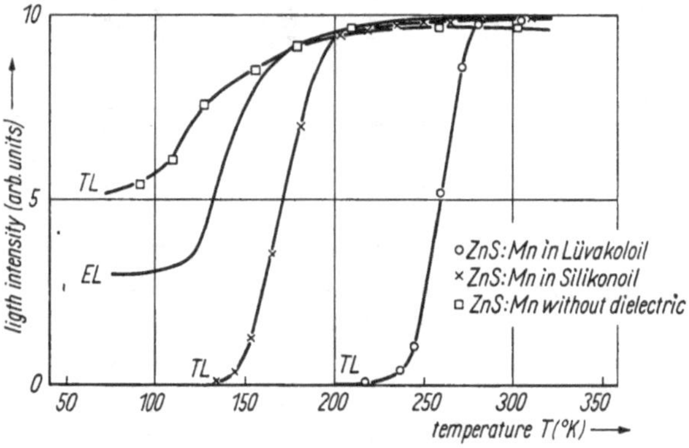

Fig. 25. Temperature dependence of TL and EL of ZnS:$5 \cdot 10^{-3}$ Mn

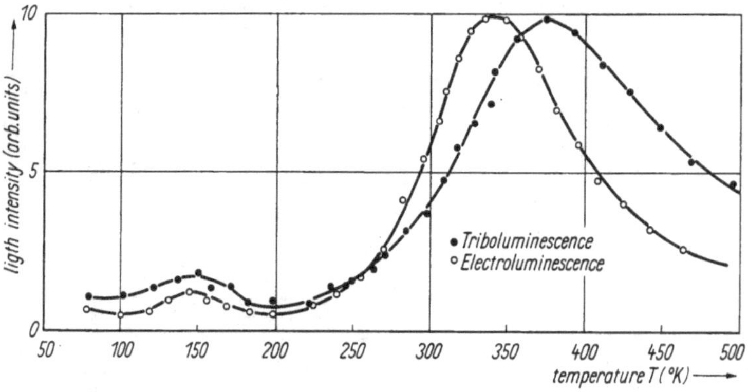

Fig. 26. Temperature dependence of TL and EL of ZnS:10^{-3} Cu, Cl

ducted due to gas discharges. The sudden decrease of the TL at 260 °K (Lüvakol oil) and 180° (silicon oil) respectively is due to the increasing viscosity of the oils at decreasing temperature. At 220 and 130 °K respectively oils has become solid and the TL-intensity decreases by the factor 200—400. The acting forces now do not only attack on the contact points between the ZnS-crystals and the glass surface but act also on the hardened interspaces. In the proportion the contact surface increases, the stress acting on the ZnS grains is bound to decrease. This reduces the TL-intensity considerably.

The temperature dependence of the TL and EL on the ZnS:10^{-3} Cu, Cl and ZnS:10^{-3} Cu, J is shown on Figs. 26 and 27. A small intensity maximum in the range 150—170 °K and a greater one between 340—410 °K occurs in both luminophores and kinds of luminescence. As with the EL these emission peaks, according to Johnson, Piper, Williams and Zalm, must be attributed to the effect of traps, the temperature dependence of TL must be attributed to the same causes in view of the similarity between TL and EL.

The temperature dependence of TL and EL (taken from a study of GOBRECHT, HAHN, GUMLICH) on the ZnS:Ag, as shown on Figure 28, also expresses the influence of the traps on EL and tribo-induced electroluminescence.

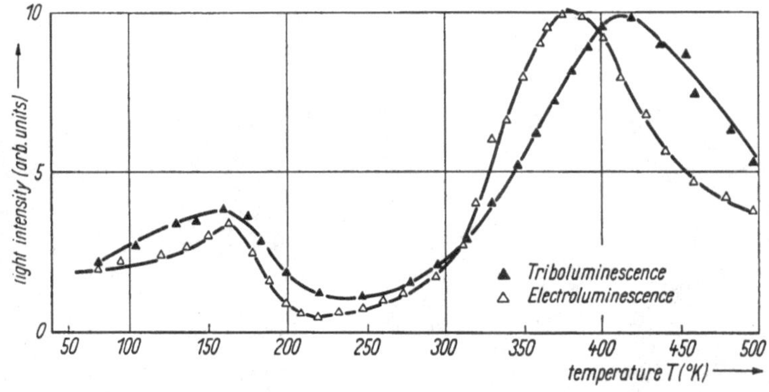

Fig. 27. Temperature dependence of TL and EL of ZnS:10^{-3} Cu, J

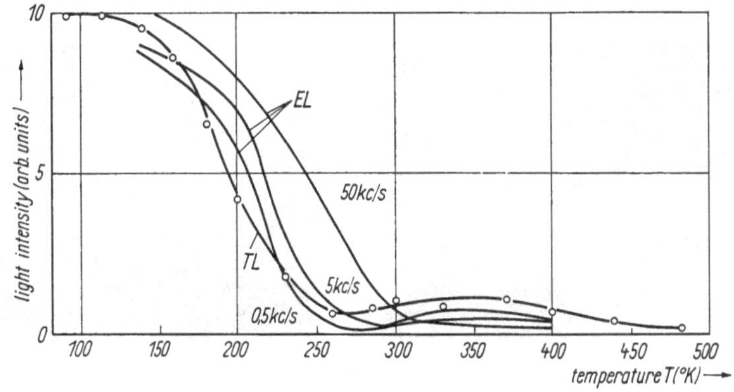

Fig. 28. Temperature dependence of TL and EL of ZnS:Ag

3.1.5 Influence of the Kind of Deformation on the TL — Periodic TL

The correlation between elastic-plastic deformations, fracture, and the development of light was investigated by KÜRTEN, RUMPF and CHUDÁČEK, SODOMKA. Contrary measuring results were to be observed. Contrary to CHUDÁČEK and SODOMKA, KÜRTEN, RUMPF found TL in ZnS:Mn only if the crystals fractured. It can be taken for granted that this discrepancy is to be attributed to the different deformation velocities, although the latter authors do not give any details on that. The deformation velocity of the compression device used by KÜRTEN was extremely low in the elastic-plastic region because the crystals were loaded quasi-statically. They reached higher values only just before the fracture. The small time resolution prevented an accurate time assignment of fracture — light within $< 10^{-5}$ s. Our own investigations indicate the great importance of the deformation velocity for the TL light intensity during the compressive treatment. At decreasing values the deformation velocity moves towards zero, alhough the amount of light remains constant in the first order approximation. This is the cause for the fact that TL never occurs at constant load but only at sufficiently quick changes of the pressure. With slow or constant deformations a polarization field develops in analogy to the EL in powders. This opposed to the causal field and compensates the same. Excitation and emission processes are thus impossible.

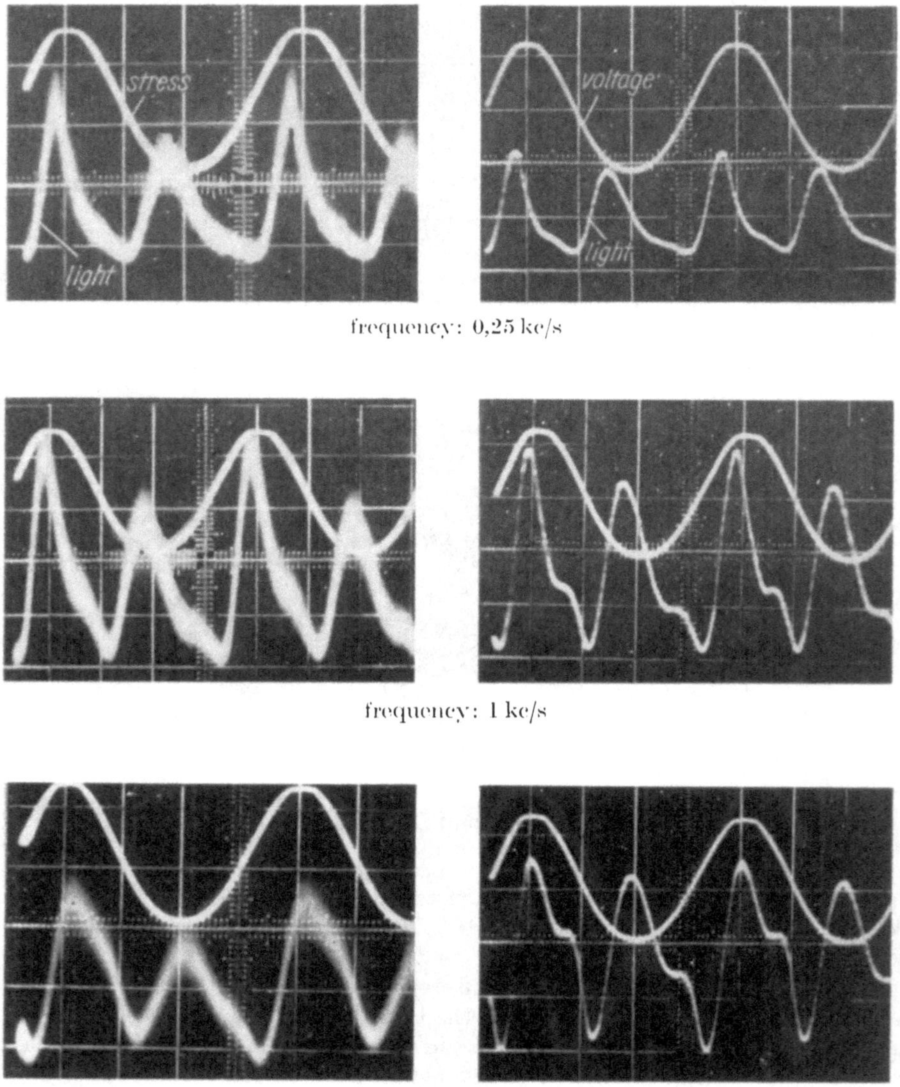

frequency: 0,25 kc/s

frequency: 1 kc/s

frequency: 5 kc/s

Fig. 29. Brightness waves of TL and EL at sinusoidal excitation (ZnS: 10^{-3} Cu, Cl). (Left-TL-brightness waves, right-EL-brightness waves)

In the light of these findings, the experiments were conducted at higher deformation velocities. To prevent pressure peaks, single crystals (ZnS:Mn) of the dimensions 1 × 1 × 1 mm were bonded between two plane surfaces and treated with compression pulses. The course of the stresses was followed up in parallel with the light. At 1/20 of the break stress of the ZnS light could be detected without the crystals having changed in any way when examined microscopically. Thus it has been proved that TL is possible in the elastic-plastic range of the ZnS.

Periodical TL, i.e. a mechanical light generator, was produced to support this result experimentally. Mechanical vibrations were produced by applying AC voltages (300 V) to a piezo-ceramics. The same were transferred to the ZnS:Cu-layer of 100 μm thickness located in the immediate neighbourhood. The stresses were sufficient as to generate TL. At a frequency of 10 kc/s, practically no TL-decrease could be observed. Within 20 min. this corresponds to a constancy after 10^7 mechanical treatment pulses. It can be concluded, therefore, that elastic deformations of the ZnS alone are sufficient to generate TL.

The TL and EL brightness waves at sinusoidal excitation shown on Figure 28 are identical. They can be interpreted by means of the delayed recombination, e.g. according to PATEK. Figure 30 shows TL and EL pulses at rectangular excitation. We can see that only changes of the stress or voltage result in a light emission.

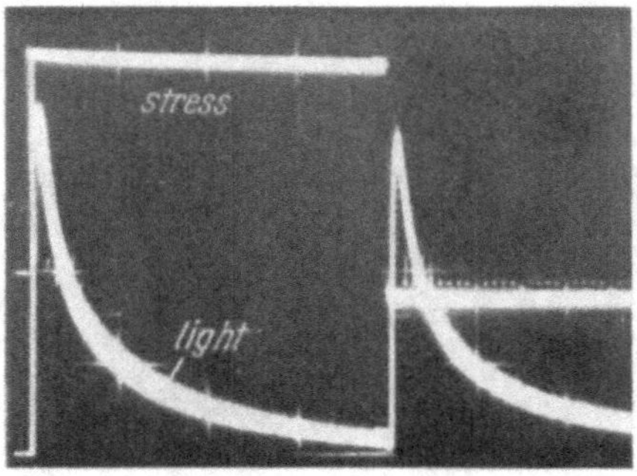

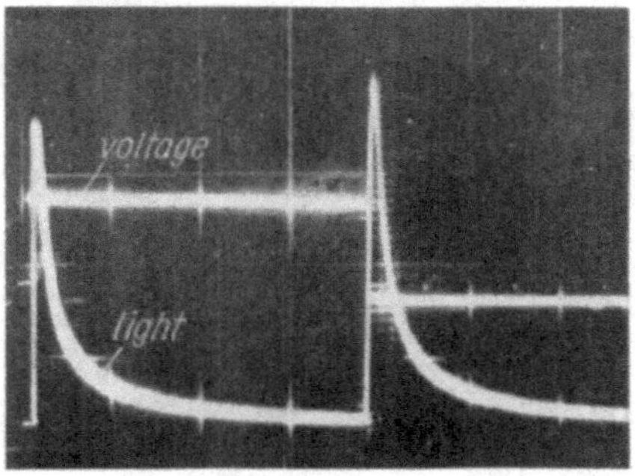

Fig. 30. Light pulses of TL and EL at rectangular excitation ($ZnS: 10^{-3}$ Cu, Cl). (Left-TL-pulses, right-EL-pulses)

3.1.6 Tribo-Induced Photoelectroluminescence (PEL)

By photoelectroluminescence (PEL) we understand an effect found on a ZnS:Mn, Cl — layer by CUSANO, i.e. a light emission under conditions of an applied electrical field was detected only if additional light was irradiated on the phosphor. Under conditions of a sole EL-excitation, the phosphor did not show any EL-emission and at a slight sole UV-excitation a very weak light emission was obtained.

Only by the simultaneous effect of the field and the irradiation with light it was possible to substantially increase the luminescent emission. The irradiation with light obviously has a stimulating effect and creates the conditions for initiating the EL-electron processes.

In complete analogy to this effect, we could prove a TL-emission in the periodic treatment of the $ZnS:5\cdot10^{-3}$ Mn only if a little additional light was irra-

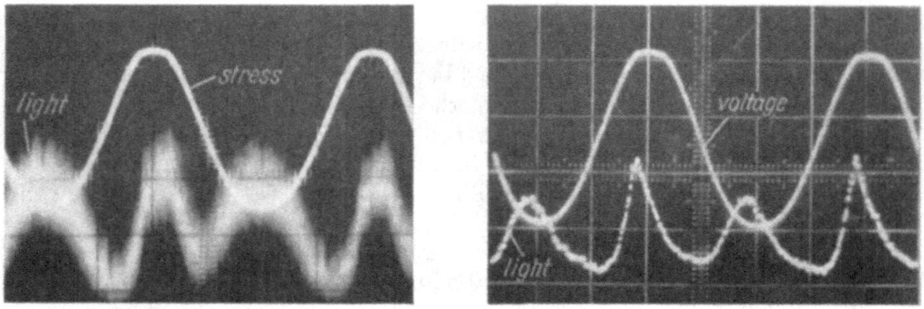

frequency: 0,5 kc/s

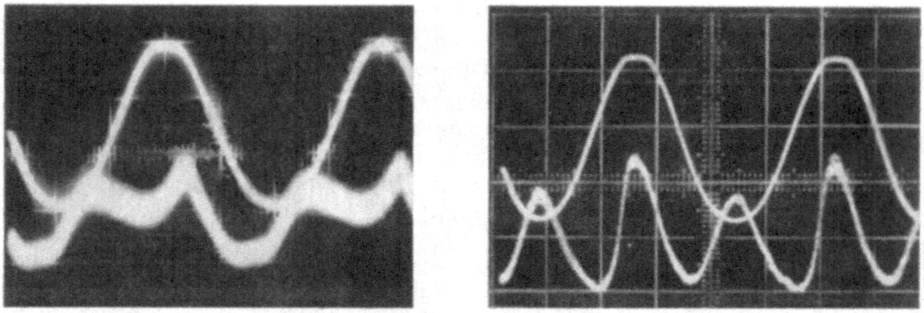

frequency: 5 kc/s

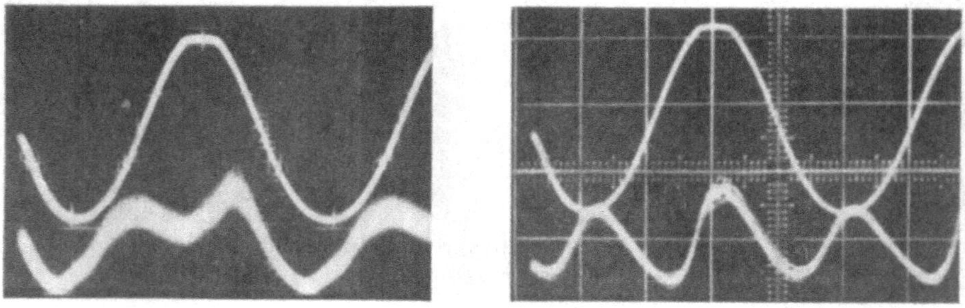

frequency: 10 kc/s

Fig. 31. TL-brightness waves with irradiation of light and EL-brightness waves without irradiation of ZnS: $5 \cdot 10^{-3}$ Mn
(Left-TL-brightness waves with irradiation, right-EL-brightness waves without irradiation)

diated. The intensification of the TL-emission (approx. 100-fold) increases considerably with the irradiated light intensity and the frequency of the mechanical vibrations. The TL is modulated in the rhythm of the stresses Figure 31 shows the TL brightness waves at simultaneous irradiation with light, and the EL brightness waves. The shift of the TL-maxima as compared with those of the EL is a consequence of the modified internal field in the phosphor. Beside the shift mentioned, we can see the slower decay of the TL as compared with the EL.

3.2 Theory

The TL-spectra, the stress-light-correlation, the dependence of the TL-intensity on the acting stresses, the decay behaviour, and the temperature dependence prove that the TL of the zinc sulphides is to be considered as a tribo-induced electroluminescence. The electric fields required for the electrical excitation of the TL are bild up during the treatment of the ZnS by charge transitions due to different contact potentials (discussed under 4.2.1), by the piezoelectricity of the zinc sulphide, by the migration of electrically charged dislocations according to Merten, and by the formation of charged fractures (e.g. 111). Owing to the different kinds of a possible field generation in the ZnS TL could be possible in the case of contact potential at pure contact without any deformation, piezoelectricity at elastic deformation, charged dislocations at plastic deformation, and charged fracture surfaces. By way of experimental investigations in dependence on the deformation it was possible to prove very many variants with the exception of the first one (pure contact).

3.2.1 TL at Elastic and Plastic Deformation

Most of the investigations conducted under 3.1 were carried out during an elastic-plastic deformation. From the direct proportionality between the force K measured during the treatment and the triboinduced voltage U causing the luminescence,

$$K = \alpha \cdot U$$

with the proportionality constant as per 3.1.2.1 being determined by the measurement of b and B, follows that the dominating cause of the tribo-induced voltage U must be attributed to the piezoelectricity of the hexagonal and cubic zinc sulphide. Thus, the direction of the electric field in the crystal is completely independent on the kind of treatment and the location of the point of attack of the stress, whereas the sign of the electric field vector follows from the dilatation or compression.

A correlation between α and the piezo compression constant g can be established on the basis of the piezoelectric field generation. To simplify the calculations we shall consider only one coordinate and one deformation perpendicular to the cubic (111)-surface. Under presence of the stress, the electric field is generated in the piezoelectric solid.

$$\mathfrak{E} = \sigma \cdot g \qquad \text{and the voltage} \tag{6}$$

$$U = \sigma \cdot d \cdot g\,, \tag{7}$$

if d is the crystal thickness in the direction considered. From equating the equations (7) and (3) follows

$$\sigma = \frac{K}{\alpha \cdot d \cdot g} = \frac{K \cdot B^2}{b^2 \cdot dg} \tag{8}$$

On the other hand, according to the definition of the stress, there exists

$$\sigma = \frac{K}{F_K} \tag{9}$$

with F_K being the contact surface betwen crystallite and glass plate, i.e. the surface on which the force K attacks directly. From the equations (8) and (9) we obtain

$$F_K = \alpha \cdot d \cdot g = \frac{b^2}{B^2} d\, g \tag{10}$$

and the possibility of determining the real contact surface from the measurement of the rise constants b and B. The quotient between F_K and the externally visible contact surface F was found to be 1:54, 1:60, and 1:74 für ZnS:10^{-3} Cu, Cl; ZnS:10^{-3} Cu, J, and ZnS:$5 \cdot 10^{-3}$ Mn.

This implies that only about 1/60 of the total surface of ZnS-powder was in contact with the treating glass plate.

The equations (1), (5) and (10) now make it possible for us to express the intensity of the tribo-induced electroluminescence as a function of the force K available for the treatment, the real contact surface F_K, the EL rise constant B, and the maximum EL-intensity I_0:

$$I_{TL} = I_0 \frac{F_K}{F} e^{-\frac{B\sqrt{F_K}}{\sqrt{dg} \cdot \sqrt{K}}}. \tag{11}$$

From this equation we can see that the correlation (1) obtained experimentally can occur only if the real contact area F_K and thus the quotient $\mu = F_K/F$ remain constant at treatment with various forces K. If that fails to be so, i.e. if μ were a function of K, the TL-intensity, measured off over lg $K^{-1/2}$, could not result in a straight line. For the special conditions $\mu \sim K^{1/2}$, $\mu \sim K^{3/4}$, and $\mu \sim K$ the curves following from the equation (11) were calculated and plotted on Figure 32. By means of these curves, the deviation from the straight lines at ZnS:$5 \cdot 10^{-3}$ Mn shown in Figure 13 for small K may be explained. $\mu = F_K/F$

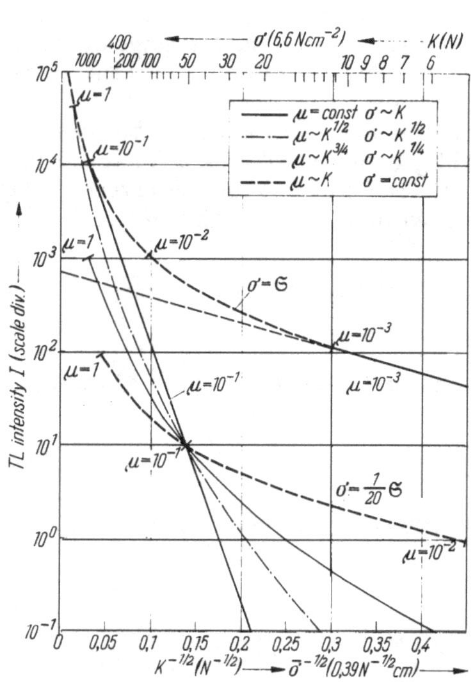

Fig. 32. Intensity of TL as a function of force K and the mean stress $\bar{\sigma} = K/F$ for $\mu = 10^{-3}$, 10^{-1}, $\mu \sim K^{1/2}$, $\mu \sim K^{3/4}$ and $\mu \sim K$ (calculated)

here increased at increasing force K. On the other hand, Figure 32 also proves that the treatment in the elastic-plastic region remained in that region and that there was little fracture. If that were not so there would have to be a crushing as from a certain force of treatment K, i.e. there would be an increase of μ and the transition from the straight line to the hyperbola ($\mu \sim K$).

For an illustration of the relationships, we shall now assess the electric fields $\mathfrak{E}$ at action of force. The stresses σ at compressive excitation were found to be between 1/30 and 1/3 of the break stress of the ZnS. At a compressive constant for ZnS (blende) of $g \simeq 4.5\ \mathrm{V\ N^{-1}}$ we obtain a field between 10^4 and $10^5\ \mathrm{V cm^{-1}}$. In conformity with 3.1.2 this is absolutely sufficient in order to generate an intensive TL with the electroluminescent substances. Although non-electroluminescent powder show a smaller TL-intensity (table 2), they ought to be still smaller as shown by comparative measurements. The discrepancy can be overcome by means of the model of the local piezoelectric properties of the luminescent centres as proposed by CHUDÁČEK (1967).

3.2.2 TL — Intensity Differences between ZnS:Mn and ZnS:Cu

The strong TL-intensity of ZnS:Mn as compared with ZnS:Cu and other dopings can be interpreted on the basis of equation (11) and the TL-decay. First of all, the cause is to be found in the greater EL-constant B, as compared with ZnS:Cu, and in the 50 times greater decay time constant. Both factors together are responsible for the strong intensity differences (200:1).

3.3 Technical Application

The TL offers the possibility of gaining general information on physical processes in the solid because it links up electrical, optical and mechanical phenomena. Technical applications have become known only recently. In this connection we should like to refer especially to the studies of Rumpf. He was able to show that the TL-effect in the crushing technique can solve important problems. By means of ZnS:Mn it was possible for the first time to understand the

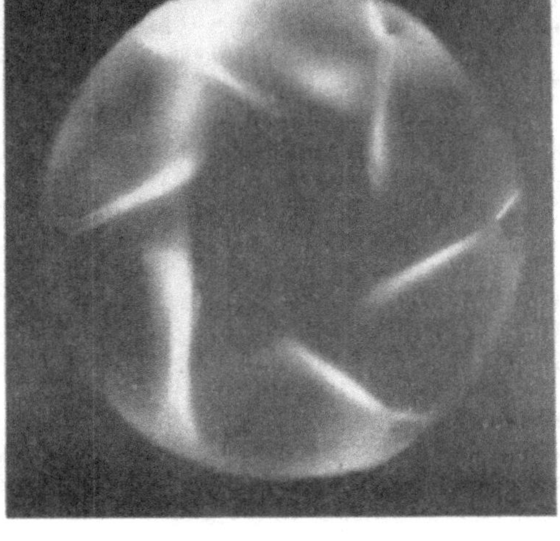

Fig. 33. Light emission in the jet mill

method of operation of a spiral jet mill and to supply optimum conditions for the construction and setting of such mills. By the use of triboluminescent-active substances as grinding substance, it was possible for the sites of the maximum crushing intensity to be made directly visible, because they are at the same time the sites of the strongest TL-emission. The luminous phenomenon generated in the spiral jet mill is shown on Figure 33. We can see that the size-reduction is not due to collisions between grains and mill walls but mainly to collisions of grain-grain. Beside this general statement on the crushing intensity, the gradation of colours in the mill described under 3.1.1.2 can be consulted for more precise description of the Size-reduction processes taking place. It is possible to assess the particle velocity, the times for the whereabouts of a grain in the mill, and the frequency of treatment a grain is subjected to.

Further applications of the TL, in particular of that the ZnS:Cu, are to be expected in view of the TL stress dependence and the dependence of the TL on the mechanical pre-history of the phosphor.

References

Alzetta, G., Minnaja, N., Santucci, S.: Il Nuovo Cimento **23**, 910 (1962)

Alzetta, G., Chella, G., Santucci, S.: Physics Letters [Amsterdam] **26 A**, 94 (1967)

Beals, C. S.: Proc. Trans. Roy. Soc. Canada **17**, Sect. III, 125 (1923)

Bersis, D. S., Nikokavouras, J.: Z. Phys. Chem. N. F. **65**, 147 (1969)

Broser, I., Gumlich, H.-E., Moser, R.: Z. Naturforsch. **20a**, 1648 (1965)

Bowden, F. P., Tabor, D.: The Friction and Lubrication of Solids, Part II, Oxford 1964

Burke, I.: Chem. News **78**, 156 (1898) Nature (London) **58**, 533 (1898)

Chudáček, J., Sodomka, L.: Čzech. J. Physics **B 13**, 209 (1963)

Chudáček, J.: Čzech. J. Physics **B 16**, 520 (1966)

Chudáček, J.: Čzech. J. Physics B **17**, 34 (1967)

Coleburn, N. L., Solow, M., Wiley, R. C.: J. appl. Physics **36**, 507 (1965)

Cusano, D. A.: Physic. Rev. **98**, 546 (1955)

Era, K., Shionoya, S., Washizawa, Y.: J. physic. Chem. Solids **29**, 1827 (1968)

Gobrecht, H., Hahn, D., Gumlich, H.-E.: Z. Physik **136**, 623 (1954)

Gross, G., Stranski, I. N., Wolff, G.: Z. Elektrochem., Ber. Bunsenges. physik. Chem. **59**, 341, 346 (1955)

Inoue, T., Kunitomi, M., Shibata, E.: J. Sci. Hiroshima Univ. **A 9**, 129 (1939) J. Chem. Soc. Japan **60**, 149 (1939)

Jenny, D. A.: J. appl. Physics **28**, 1515 (1957)

Johnson, P. D., Piper, W. W., Williams, F. E.: J. electrochem. Soc. **103**, 221 (1953)

Knyasewa, N. P.: see Reference Lobachew

Koda, T., Shionoya, S., Ichikowa, M., Minomura, S.: J. physic. Chem. Solids **27**, 1577 (1966)

Krotowa, N. A.: see reference Lobachew

Kürten, H., Rumpf, H.: Kristall u. Technik **1**, 627 (1966)

Laisaar, A. J., Kirs, J. J.: Ber. d. Ind. f. Phys. u. Astron. d. Ak. d. W. der Estn. SSR **18**, 23 (1962)

Linke, E.: Z. angew. Physik (presented)

Lobachew, A. N., Kusmina, N. P., Martyshew, Jn. N.: Paper presented to the Soviet Conference on Mechano-Emission and Mechano-Chemistry in Frunse, June 23—30, 1969 (be published in a special conference volume)

LONGCHAMBON, H.: C. R. Séances Acad. Sci. **174**, 1633 (1922), **176**, 691 (1923); Bull. Soc. franc. Minéralog. **48**, 130 (1925)
MEYER, K., POLLY, F.: Physica Status solidi **8**, 441 (1965)
MEYER, K., OBRIKAT, D.: Z. physik. Chemie **240**, 309 (1969)
MEYER, K., OBRIKAT, D., ROSSBERG, M.: Physica Status solidi **35**, K 13 (1969)
MEYER, K., LINKE, E.: Physica Status solidi **26**, K 89 (1968)
MUSCHENBROCK, P. v.: Tentamina experimentorum naturalicim captorum in academia del cimento. Lugduni Batavorum H. Verbeck 1731 Pars II, p. 185
NELSON, D. M.: J. opt. Soc. America **16**, 207 (1926)
OBREIMOFF, J. W.: Proc. Roy. Soc. [London] Ser. A **127**, 294 (1930)
OBRIKAT, D., MEYER, K., POLLY, F.: Physica Status solidi **22**, K 123 (1967)
OBRIKAT, D.: Dissertation, Humboldt-Universität Berlin 1970
OBRIKAT, D.: Physica Status solidi (1970) to be published
ORTMANN, H., TREPTOW, H.: Z. Naturforsch. **16**, 910 (1961)
ORTMANN, H., TREPTOW, H.: Physica Status solidi **3**, K 13 (1963)
PRENER, J. S., WILLIAMS, F. E.: J. Physique Rad. **17**, 667 (1956)
PATEK, K.: Čzech. J. Physics **9**, 460 (1959)
POLLY, F., OBRIKAT, D., MEYER, K.: Z. angew. Physik **22**, 516 (1967)
RUFOW, Ju., N. KADUSCHIN, A. A., ROGINSKI, S.: Doklady Akad. Nauk. SSSR **171**, 905 (1966)
RUMPF, H.: Chemie-Ing.-Techn.: **31**, 323 (1959), **32**, 129 (1960), **34**, 731 (1962)
SCHÖNERT, K., KÜRTEN, H.: Triboluminescenzsymposium, Berlin 1967 (unpublished)
SODOMKA, L.: Čzech. J. Physics **B 14**, 800 (1964)
SODOMKA, L.: Čzechoslov. Časopis pro Fysika **A 18**, 497 (1968)
STRANSKI, I. N., STRAUSS, E., WOLFF, G.: Z. Elektrochem. angew. physik. Chem. **55**, 633 (1951); Z. Elektrochem., Ber. Bunsenges. physik. Chem. **59**, 341 (1955)
VAN GOOL: Philips Res. Rep., Suppl. **3**, 1 (1961)
THIESSEN, P. A., MEYER, K., HEINICKE, G.: Grundlagen der Tribochemie. Abh. Dtsch. Akad. Wiss. Berlin, Kl. Chem., Geol., Biol. 1967
WAGGONER, C. W.: Physic. Rev. **7**, 402 (1916)
WIEDEMANN, G., SCHMIDT, F.: Wiedemanns Ann. Physik **54**, 604 (1895)
WOLFF, G., GROSS, G., STRANSKI, I. N.: Z. Elektrochem. angew. physik. Chem. **56**, 420 (1952)
WOOD, R. W.: Proc. Roy. Soc. [London] Ser. A **99**, 362 (1921)
ZALM, P.: Philips Res. Rep. **11**, 353, 417 (1956)

(Received September 12, 1969)

Authors' address:

Prof. Dr. habil. K. MEYER
Fachbereich Physikalische Chemie
der Sektion Chemie der
Friedrich-Schiller-Universität Jena
69 Jena
Lessingstr. 10

Dr. D. OBRIKAT
Dipl.-Phys. M. ROSSBERG
Zentralinstitut für Physikalische Chemie
der DAW zu Berlin
1199 Berlin-Adlershof
Rudower Chaussee 5

Kristall und Technik	5	1	1970	51–59

I. P. Kalinkin, K. K. Muravyeva, L. A. Sergeyewa,
V. B. Aleskowsky, N. S. Bogomolov

Institute of Technology (Lensovieta), Leningrad

Growth and Structure of Monokristalline Films of $A^{II}B^{VI}$ Compounds

In vacuum 10^{-4}—10^{-5} torr monocrystalline thin layers of CdS, CdSe, CdTe, ZnS, ZnSe, ZnTe were prepared on mica, (111) surfaces of germanium and gallium arsenide single crystals and (0001) and (11$\bar{2}$0) cadmium sulphide surfaces. The film structures and some of their properties (conductivity, n- or p-mobilities) were shown to depend on temperature conditions of film preparation. It was shown experimentally that in the case of monocrystalline thin films of zinc and cadmium chalcogenides there is a correlation between evaporation and epitaxy temperatures of film preparation. Structure and phase composition of monocrystalline layers are connected with the nature of substrate substances used (including polarity of ⟨111⟩ direction in gallium arsenide and ⟨0001⟩ direction in cadmium sulphide).

В вакууме 10^{-4} —10^{-5} торр получены монокристаллические тонкие слои CdS, CdSe, CdTe, ZnS, ZnSe, ZnTe на слюде, поверхностях (111) монокристаллов германия и арсенида галлия, поверхностях (0001) и (11$\bar{2}$0) монокристаллов сульфида кадмия. Исследована зависимость степени совершенства структуры, фазового состава, а также некоторых свойств (удельного сопротивления и подвижности носителей заряда) пленок от температурного режима выращивания. При синтезе монокристаллических тонких слоев халькогенидов кадмия и цинка экспериментально установлена взаимозависимость между температурами испарения и эпитаксии. Показано влияние природы ориентирующих подложек (и, в том числе, полярности осей ⟨111⟩ арсенида галлия и ⟨0001⟩ сульфида кадмия), а также влияние конструкции испарителя на совершенство структуры и фазовый состав монокристаллических пленок.

With the development of semiconductor techniques, microelectronic and acustic the question of synthesis of monocrystalline films of A^{II} B^{VI} compounds has been given ever increasing attention. However, no systematic studies on the growing of the above films as well as investigation of their structure and electrophysical properties are reported in the literature.

The purpose of the present paper was to study

a) one-, two- and three layers of zinc and cadmium chalcogenide films on various orienting substrates (mica — muscovite, germanium (111), gallium arsenide {111}, monocrystals of CdS (0001) and (11$\bar{2}$0); PbS_k, $PbSe_k$ monocrystalline films,
b) epitaxial growth of A^{IV} B^{VI} films on monocrystalline A^{II} B^{VI} films,
c) the relation of the degree of perfection of epitaxial CdS, CdSe, CdTe, ZnS, ZnSe, ZnTe films on mica to the epitaxial temperature (T_{ep}) and evaporation temperature (T_{ev}) — that is the degree of supersaturation,
d) the specific resistance (ϱ) and current carriers mobilities (μ) in films condensed on mica.

Table 1

One and multi-layer monocrystalline chalcogenide $A^{II}B^{VI}$ and $A^{IV}B^{VI}$ films obtained by sublimation in vacuum $10^{-4} - 5 \cdot 10^{-5}$ torr onto orienting substrates

One-layer films

(CdS_h) $CdSe_h$; $ZnTe_c$)
↓
Ge (111) GaAs (111) GaAs ($\bar{1}\bar{1}\bar{1}$) mica CdS (0001) CdS (11$\bar{2}$0)

$CdSe_c$
↓
Ge (111) GaAs ($\bar{1}\bar{1}\bar{1}$) mica

$CdSe_{c+h}$
↓
Ge (111) GaAs (111) GaAs ($\bar{1}\bar{1}\bar{1}$) mica CdS (0001)

$CdTe_c$
↓
Ge (111) GaAs (111) mica CdS (11$\bar{2}$0)

ZnS_c
↓
Ge (111) mica CdS (0001)

GaAs ($\bar{1}\bar{1}\bar{1}$) ← ZnS_h → mica

$ZnSe_c$
↓
Ge (111) GaAs (111) GaAs ($\bar{1}\bar{1}\bar{1}$) mica

Ge (111) ← $ZnSe_h$ → mica

two-layer films

$A^{II}B^{VI}$ (CdS_h; $CdSe_h$; $CdTe_c$; $ZnSe_c$; $ZnTe_c$)
↓
CdS → mica $CdSe_h$ → mica $CdTe_c$ → mica
↑
$A^{IV}B^{VI}$ (PbS_c, $PbSe_c$)

three-layer films

mica ← CdS_h ← $PbSe_c$ ← CdS_h
mica ← $CdTe_c$ ← $PbSe_c$ ← CdS_h
mica ← $CdTe_c$ ← PbS_c ← CdS_h

[1]) Letters (h) and (c) denote hexagonal and cubic structures respectively, the arrows show on what kind of orienting substrates (either $A^{II}B^{VI}$ or $A^{IV}B^{VI}$ compounds) have been spread. The thickness of films was 0,05 – 1 μm. The temperature of substrates during condensation was 250 – 350 °C. ($\bar{1}\bar{1}\bar{1}$)-surface, Ga{111} and (111)-sufrace As{111}.

1. Experimental

The starting cadmium, zinc and lead chalcogenide powders were of "semiconductor" grade. ZnS powder for "luminophores" was preliminary calcined in a current of $H_2 + H_2S$ at 1100 °C. The film condensation was carried out in a vacuum of 10^{-4} to 10^{-5} torr. A quartz ampoule of the Knudsen cell type (inner diameter about 6 mm and that of the capillary 0.6—0.8 mm) was used as an evaporator, the distance from the evaporator to the substrate being 85—100 mm.

Gr (111), GaAs {111}, CdS (0001) orienting substrate plates have been chemically etched beforehand (Kalinkin et al.).

2. Results

Table 1 and Figures 1a, 1b give the results of $A^{II} B^{VI}$ film growth investigation which show that at relatively low epitaxial temperatures (250—350 °C) monocrystalline films of these compounds may be synthesized on substrates of Ge (111), GaAs {111}, on mica (muscovite), monocrystals of CdS (0001); on PbS, PbSe monocrystalline films by means of vacuum condensation.

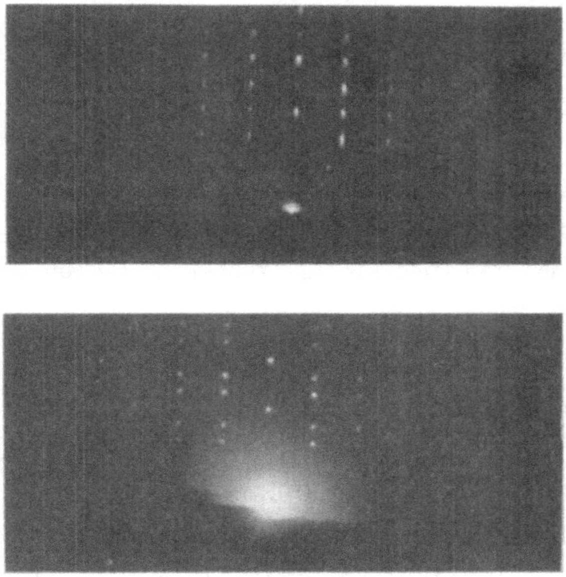

a)

b)

Fig. 1. Electronograms of monocrystalline films
a) $CdS_{(h)} \rightarrow GaAs_{(c)}$ $(\bar{1}\bar{1}\bar{1})$, b) $CdSe_{(c)} \rightarrow GaAs$ $(\bar{1}\bar{1}\bar{1})$

It follows from Table 1 that monocrystalline $A^{II}B^{VI}$ film growth of predetermined structure (cubic, hexagonal) on $A^{III}B^{V}$, A^{II} B^{VI} monocrystals may obviously be easier achieved, if one takes into consideration the polarity of substrate planes {111}. For example, CdSe films condensed under similar conditions on GaAs $\{\bar{1}\bar{1}\bar{1}\}$-planes crystallize in cubic modification on Ga (111)-planes and in hexagonal modification on As (111)-planes.

The investigation of $A^{II}B^{VI}$ epitaxial films on mica made it possible to state the relation between the epitaxial temperatures (T_{ep}) and evaporation temperatures (T_{ev}) at which monocrystalline cadmium (Fig. 2a) and zinc (Fig. 2b) chalcogenide films are growing.

Figures 2a, 2b show that the relation $T_{ev} = f(T_{ep})$ for all compounds is at its maximum in the temperature region 310—320 °C.

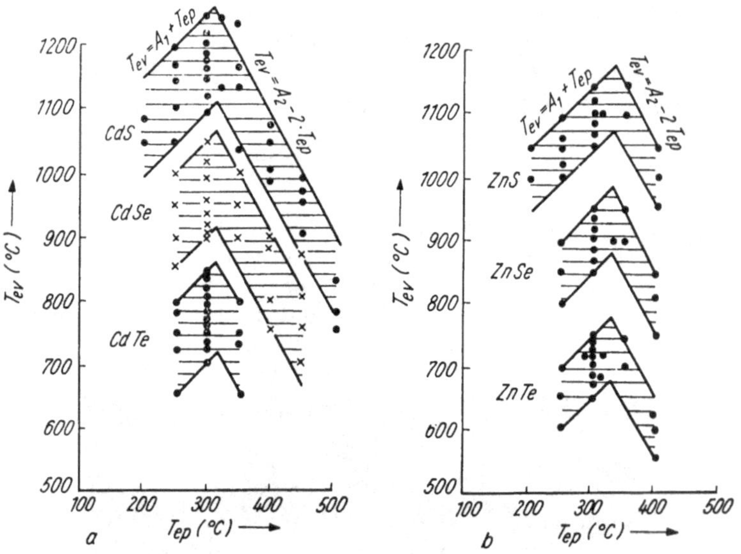

Fig. 2. Relation of evaporation temperature T_{ev} to that of epitaxy T_{ep} when growing monocrystalline cadmium (a) and zinc (b) chalcogenide films on mica

Tev for each of the chalcogenides is approximated by the equations

$$T_{ev} = A_1 + T_{ep} \quad (1)$$

$$T_{ev} = A_2 - 2\,T_{ep} \quad (2)$$

where A_1 and A_2 are constants calculated from the graphs. They are given in Tables 2 and 3. Graphs 2a, 2b and equation (2) made it possible to estimate the critical temperature of monocrystalline film growth T_{en} (crit.) (see Tables 2, 3). However, practically these values proved to be lower than those calculated; the tendency of monocrystalline films to grow at high epitaxial temperatures decreased. This seemed to be due to considerable gas evolution from mica and to the increased action of residual gases on the structure of films growing at low rates. The character of the relation $T_{ev} = f(T_{ep})$ can be explained by comparing the data on thermal, thermographical and mass — sectrometric studies of mica gas separation (LASHEV, ZVETKOV; LANGMUIR; WAGENER, MARTH; MERZER). Mass-spectrometric data (MERZER) show that at 200 °C there occurs a slight gas evolution from the surface of mica: nitrogen (26%) and CO (67%) but scarcely water (3%). The screening effect of both absorbed water and gases at that temperature of the substrate hinders the epitaxial growth of $A^{II}B^{VI}$ monocrystalline films. Within the temperature range from 200 to 300 °C there is a gradual cleaning of the surface and the total amount of desorbed gases increases (from $\Sigma p = 8{,}3 \cdot 10^{-8}$ to $\Sigma p = 3{,}2 \cdot 10^{-6}$ torr) and their quantitative ration changes (50% CO; 22% N_2;

Table 2

Growth rate of monocrystalline ZnS, ZnSe, ZnTe films. Degree of supersaturation (α) close to the substrate surface. (A_1, A_2 constants)

Epitaxial Temperature[3]) T_{ep} (°C)	ZnS Films growth rate Å · sec^{-1}	ZnS Films degree of supersaturation[2]) $\alpha = P/P_0$	ZnS Films lg K_{ZnS}	ZnS Films	ZnSe Films growth rate Å · sec^{-1}	ZnSe Films degree of supersaturation $\alpha = P/P_0$	ZnSe Films lg K_{ZnSe}	ZnSe Films	ZnTe Films growth rate Å · sec^{-1}	ZnTe Films degree of supersaturation $\alpha = P/P_0$	ZnTe Films lg K_{ZnTe}	ZnTe Films
200[1])	1—10	$3.7 \cdot 10^{11} - 3.7 \cdot 10^{12}$	$\lg K_{ZnS} = -\frac{41\,280}{T} + 22.7$	$A_2 = 1750-1850$ $A_1 \approx 750-850$	0.5—1	$(1.2-2.4) \cdot 10^{8}$	$\lg K_{ZnS} - \frac{38\,400}{T} + 21.2$	$A_2 \approx 1550-1650$ $A_1 \approx 550-650$	1	$2.7 \cdot 10^{6}$	$\lg K_{Zne} - \frac{33\,920}{T} + 21.2$	$A_2 \approx 1350-1450$ $A_1 \approx 350-450$
250	1.5—22.5	$(0.9-1.2) \cdot 10^{9}$			1—2	$(3.0-6.0) \cdot 10^{7}$			0.8—5	$1.5-10^{4} - 1.0 \cdot 10^{5}$		
300	8—14	$(2.6-4.5) \cdot 10^{7}$				$6.1 \cdot 10^{-5} - 1.6 \cdot 10^{6}$			1—4	$3.4-10^{2} - 1.4 \cdot 10^{3}$		
350	0.9—3	$3.8 \cdot 10^{4} - 1.3 \cdot 10^{5}$			1.5—3	$6.0 \cdot 10^{3} - 1.2 \cdot 10^{4}$			1—2	$1.1 \cdot 10^{1} - 2.2 \cdot 10^{1}$		
400	1—1.7	$(0.9-1.6) \cdot 10^{3}$			2—4	$2.5 \cdot 10 - 6.0 \cdot 10$				0.2—2.8		

[1]) At $T_{ep} = 200$ °C only ZnS and CdS monocrystalline films could be grown Films of the other compounds were texturizes.

[2]) P-real vapour pressure close to the substrate at T_{ep}. The latter was calculated from the rate of film growth SHALIMOVA P_0 – equilibrium vapour pressure close to the substrate at T_{ep}; calculated from the equation P_0 (torr) $= 3/2 \cdot 2^{1/3} \cdot 760 K_{MX}^{1/3}$ LORENZ, and using logarithms follows: $\lg P_0 = 1/3 \cdot \lg K_{MX} + 3{,}15$. To Compute the values of $\lg K_{MX}$ at different T_{ep} approximate equations $\lg K_{MX} = -\frac{\Delta H^0_{298}}{4{,}576\, T} + \frac{\Delta S^0_{298}}{4{,}576}$ have been used; ΔH^0_{298} and ΔS^0_{298} for the dissociation reactions, e. g. $2\,CdS \rightarrow 2\,Cd + S_2$, being computed from data taken from GOLDFINGER, JEUNEHOMME; STULL, SINKE.

[3]) T_{ep} (crit.) CdS $= 585 - 635$ °C, T_{ep} (crit.) CdSe $= 525-575$ °C, T_{ep} (crit.) CdTe $= 450-500$ °C, T_{ep} (crit.) ZnS $= 575-625$ °C, T_{ep} (crit.) ZnSe $= 525-550$ °C, T_{ep} (crit.) ZnTe $= 450-480$ °C.

Table 3

Growth rate of monokristalline CdS, CdSe, CdTe films. Degree of suppersaturation (α) close to the substrate surface. (A_1, A_2 constants)

Epitaxial Temperature[3]) T_{ep} (°C)	CdS Films				CdSe Films				CdTe Films			
	growth rate Å · sec⁻¹	degree of supersaturation[2]) $\alpha = P/P_0$	lg K_{CdS}		growth rate Å · sec⁻¹	degree of supersaturation $\alpha = P/P_0$	lg K_{CdSe}		growth rate Å · sec⁻¹	degree of supersaturation $\alpha = P/P_0$	lg K_{CdTe}	
200[1])	86—100	$(1.4-1.6) \cdot 10^9$	$\lg K_{CdS} = -\frac{35020}{T} + 22.5$	$A_1 \approx 800-950$ $A_2 \approx 1750-1900$	10—35	$5.0 \cdot 10^7 - 1.7 \cdot 10^8$	$\lg K_{CdSe} = -\frac{33350}{T} + 20.6$	$A_1 \approx 600-750$ $A_2 \approx 1550-1700$	1,5—2	$(1.0-1.4) \cdot 10^5$	$\lg K_{CdTe} = -\frac{30750}{T} + 21.5$	$A_1 \approx 400-550$ $A_2 \approx 1350-1500$
250	121—128	$9 \cdot 10^6$			19—150	$5.3 \cdot 10^5 - 4.2 \cdot 10^6$			3.3—20	$9.5 \cdot 10^2 - 5.8 \cdot 10^3$		
300	117—244	$(1 \div 2) \cdot 10^5$			31—104	$1.3 \cdot 10^4 - 5.7 \cdot 10^4$			6—15	$3.5 \cdot 10^1 - 8.9 \cdot 10^1$		
350	100—187	$(2.1-3.8) \cdot 10^3$			5—135	$6.7 \cdot 10^1 - 1.8 \cdot 10^3$			2—8.5	0.53—2.2		
400	112—185	$(1.0-1.6) \cdot 10^2$			4.5—52	$2.6-3 \cdot 10^1$			1	—		
450	61—139	3.5—8			3—23	1						
500	5—8.5				0.3—0.8							

[1]) [2]) [3]) (Footnotes see Table 2 S. 55)

11% H_2; 13% H_2O; 3% CH_4). There seem to be two causes for the existence of maxima on Figures 2a, 2b. First at the temperature of 300 °C one often observed water evolution maxima (ZVETKOV) on the dehydration curves of muscovite and, hence one can expect to obtain a very near "pure" mica surface. According to LANGMUIR when calcining muscovite for three hours in a vacuum the amount of water and CO_2 removed corresponded to two layers of H_2O molecules and to one layers of CO_2, which covered the whole surface.

Secondly at temperatures 300—360 °C (depending on the nature of mica) recrystallization caused rearrangement of the points of breakage and changes in the crystalline lattice deformed while being cleaved. It is attended by an exothermic effect on the thermographs whereas the mica weight ist not changed (MERZER). The maximum position of exoeffect for mica of korelian origin we used in our experiments was observed at about 300—340 °C (ZVETKOV) (depending on the fineness of the damples), which coincides fairly well with the maximum position on Figures 2a, 2b. The above suggestions are confirmed by the relation $T_{ev} = f(T_{ep})$ at $T_{ep} > 320$ °C. In fact, at these temperatures the screenig effect of remaining gases is found to decrease considerably on the more perfect and cleaned mica surface; monocristalline films grow at much lower T_{ev} and with the less degree of supersaturation (see Tables 2, 3).

In the investigations of cadmium selenide monocrystalline films grown at an epitaxial temperature of 300 °C and at the evaporation temperature 900—950 °C the specific resistance (ϱ) of thin (0,05—0,25 μm) and thick (0,3—5 μm) films is of the same order — a proof that the composition of the film layers remains unchanged while growing. Wheras the current carriers mobility (μ) is one order lower in thin layers than in thick ones, thus showing the structure of film layers near the substrate being more imperfect. Further on, to find the relation of specific resistance and current carriers mobility to the nature of the chalcogenide, synthesis conditions and film structure only films thicker than 0,3 μm have been investigated.

Investigating the growth of cadmium selenide epitaxial films within a wide range of evaporation temperatures at T_{ep} = const. it was observed that the specific resistance is depending on evaporation temperature but slightly. On the contrary the current carriers mobility changes to a large extent with the evaporation temperature when T_{ep} = const. (250—300 °C).

At first with the rise of evaporation temperature the mobility value increases, being the highest in the region of the monocrystalline film growth. On further rise of T_{ev} the mobility increases but slightly.

The curves relating the electron mobility (Fig. 3a) in monocrystalline cadmium sulphide and cadmium selenide films to epitaxial temperature have two maxima at T_{ep} = 300 and 450 °C. As long as the increase of the electron mobility is not accompanied by the decrease of specific resistance with the rise of the epitaxial temperature from 200 to 300 °C, it is natural to consider this increase as being caused by the perfectioning of the monocrystalline film structure. A certain increase of the specific resistance of CdS and especially CdSe films when changing the epitaxial temperature from 200 to 300 °C seems to be connected with the gradual desorption of water vapours from the surface of mica and with the growth of films whose composition somewhat closer approaches stoichiometry than whith T_{ep} = 200 °C.

At higher substrate temperatures, e.g. at 350 °C (see Table 3) a copious evolution of hydrogen from mica was observed by mass-spectrometry (MERZER).

The interaction of hydrogen with sulphur and selenium vapours to form H_2S and H_2Se may be one of the causes of the decrease of the specific resistance of the films. Besides, when rising the epitaxial temperature from 350 to 500 °C the process of thermal etching of the films becomes evident. For dissociating CdS and CdSe compounds this process may result in an additional excess of cadmium in the films.

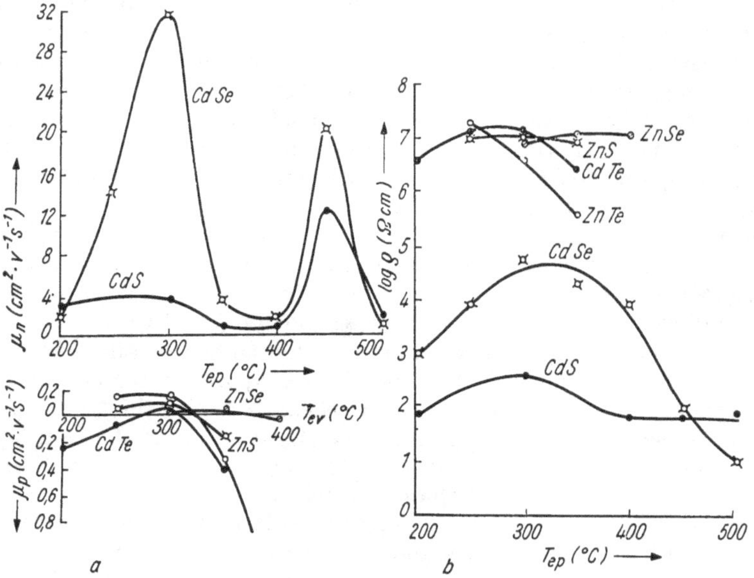

Fig. 3. Values of charge carriers mobility μ–(a) and specific resistance ϱ–(b) of monocrystalline cadmium and zinc chalcogenides obtained with different epitaxial temperatures

The mobility maximum at $T_{ep} = 450$ °C may be explained by the joint action of these factors. Thus appearance of the second electron mobility maximum does not seem to be so much due to the perfection of film structure as to an additional excess of cadmium, the decrease of CdS and CdSe film specific resistance at $T_{ep} > 350$ °C being a further proof of if.

Monocrystalline CdTe, ZnS, ZnSe, ZnTe chalcogenide films with a low growth rate had a high specific resistance and mobility inversion (Fig. 3a, b). The inversion point lies within the range of T_{ep} from 300 to 320 °C for ZnS, ZnTe, CdTe and of T_{ep} from 350 to 370 °C for ZnSe. The decrease of specific resistance value of cadmium and zinc telluride monocrystalline films which is accompanied by p-type of conductivity may be connected with the excess of tellurium atoms.

3. Summary

1. Comparison of data of Table 1 on one-, two-, and three layers of monocrystalline films leads us to assume that monocrystalline films of the type Ge – $A^{II}B^{VI}$ – $A_1^{II}B_1^{VI}$; $A^{III}B^{V}$ – $A^{II}B^{VI}$ – $A^{IV}B^{VI}$ and Ge – $A^{IV}B^{VI}$; $A^{III}B^{V}$ – $A^{IV}B^{VI}$ may also be grown on monocrystals of Ge and $A^{III}B^{V}$.

2. The relation of evaporation temperature (T_{ev}) to epitaxial temperature (T_{ep}) at which monocrystalline cadmium and zinc chalcogenide films are growing

$$T_{ev} = A_1 + T_{ep} \text{ at } T_{ep} \leqq 310\ ^\circ\text{C} \quad \text{and}$$

$$T_{ev} = A_2 - 2\,T_{ep} \text{ at } T_{ep} \geqq 320\ ^\circ\text{C}$$

has been obtained. The existence of a maximum on the curves expressing the relation $T_{ev} = f(T_{ep})$ seems to be due to recrystallization of mica and the desorption of the adsorbed gases and water from its surface at 300—320 °C.

3. It is shown that as a rule the most perfect cadmium and zinc chalcogenide films grow on mica at T_{ep} of about 300—320 °C.

4. The monocrystalline films obtained posses a great variety of properties. For example, there is a possibility to grow CdSe films both with a high and low specific resistance with a rather high electron mobility. ZnS, ZnSe, ZnTe and CdTe of n and p-type of conductivity can also be synthesized.

References

GOLDFINGER, P., JEUNEHOMME, M.: Trans. Faraday Soc. **59**, Nr. 492, Part 12, 2851 (1963)

KALINKIN, I. P., SERGEYEWA, L. A., MURAVYEVA, K. K., CHRISTOFOROW, K. K., PEVSNER, JA. G., ALESKOWSKY, V. B.: "Growth and Structure of Monocristalline Films of Semiconductors"; the works of symposium under the doctor of phis.-math. Science: "Science", Novosibirsk, part **1**, 355 (1968).

LANGMUIR, I.: J. Amer. chem. Soc. **40**, 1361 (1918).

LASHEV, E. K.: Slyuda, Moskva 1948, p. 103

LORENZ, M. R.: in Thermodynamics, Materials Preparation and Crystal Growth in Physics and Chemistry of II—VI Compounds (Ed. by M. Aven, J. S. Prener), Amsterdam 1967, p. 73

MERZER, P. D.: Vacuum **17**, 5, 267 (1967) — perevod v ekspress-informazii, Ser. Elektronika, vyp. **31**, 198 (1967)

SHALIMOVA, K. V.: Praktikum po poluprovodnikam i poluprovodnikovym priboram

STULL, D. R., SINKE, G. C.: Thermodynamic properties of the elements, New York 1956

WAGENER, J. S., MARTH, P. T.: J. appl. Physics **28**, 9, 1027 (1957)

ZVETKOV, A. I., VAL'YASHICHINA, E. P.: Materialy po termicheskomu issledovaniyu mineralov, III, Slyudy, Trudy in-ta geologii rudnych mestorojdeniy, petrografii, mineralogii i geochimii, vyp. 4, Moskva 1956, p. 13, 28, 37

(Received April 28, 1969)

Authors' address:

Prof. Dr. V. B. ALESKOWSKY
Dr. I. P. KALINKIN
K. K. MURAVYEVA
Dr. L. A. SERGEYEVA
N. S. BOGOMOLOV
Institute of Technology (Lensovieta)
Leningrad L-13, Sagorodnyj Prospect 49
USSR

Kristall und Technik	5	1	1970	61–72

L. A. Sergeyewa, I. P. Kalinkin, V. B. Aleskovsky,
A. P. Nechiporenko

Institute of Technology (Lensovieta), Leningrad

Epitaxial Growth of Oxycompounds on the Surface of Single crystalline Films and Single Crystals of $A^{II}B^{VI}$

The results of electron diffraction investigation of the composition and structure of air annealed single crystalline films and single crystals $A^{II}B^{VI}$ (CdS, CdSe, CdTe) are reported.

$A^{II}B^{VI}$ single crystalline films were grown by deposition from vapour on orienting substrates — either on the mica plane (001) or the NaCl (111) plane — with wurtzite structure (CdS_{H}, $CdSe_{H}$), sphalerite (CdS_{s}, $CdSe_{s}$, $CdTe_{s}$) or that of rock salt (CdS_{NaCl}). The lattice dimensions a of CdS_{NaCl} single crystalline films grown on mica by epitaxial growth method were found to be 5.6 Å.

It was shown that single crystalline layers of oxycompounds part of which possess formerly unknown modifications of the hexagonal structure with the parameters

$$a = 3.3 \text{ Å} \qquad c = 5.25 \text{ Å}$$
$$a = 5.5 \text{ Å} \qquad c = 7.0 \text{ Å}$$
$$a = 5.3 \text{ Å} \qquad c = 7.0 \text{ Å}$$

were formed on the surface of layers of CdS when the latter were annealed.

Chemical analysis of annealed CdS films enabled single crystalline substances with lattice dimensions $a = 3.3$ Å, $c = 5.25$ Å to be identified as polymorphous hexagonal cadmium oxide modification.

Epitaxial layers of CdO_{hex} and CdO_{NaCl} were also observed on single crystals of CdS after annealing in air.

In order to find out the mechanism of oxycompounds growth the effect of certain factors on the structure, composition and electrophysical properties (specific resistance ϱ, mobility of charge carriers μ) of initial and annealed films was investigated.

В работе описаны результаты электронографического исследования состава и структуры отожженных на воздухе монокристаллических пленок и монокристаллов $A^{II}B^{VI}$ (CdS, CdSe CdTe).

Монокристаллические пленки $A^{II}B^{VI}$ были выращены методом вакуумной конденсации на ориентирующих подложках — на плоскости (001) слюды или плоскости (111) NaCl — и имели структуру вюртцита ($CdS_{в}$, $CdSe_{в}$), сфалерита ($CdS_{сф.}$, $CdSe_{сф.}$, $CdTe_{сф.}$) либо каменной соли (CdS_{NaCl}). Параметр решетки a пленок CdS_{NaCl}, выращенных эпитаксиально на слюде, равен 5,6 Å.

Показано, что на поверхности пленок CdS при отжиге на воздухе образуются монокристаллические слои окисных соединений, часть из которых имеет ранее неизвестные модификации гексагональной структуры с параметрами

$$a = 3{,}3 \text{ Å} \qquad c = 5{,}25 \text{ Å}$$
$$a = 5{,}5 \text{ Å} \qquad c = 7{,}0 \text{ Å}$$
$$a = 5{,}3 \text{ Å} \qquad c = 7{,}0 \text{ Å}$$

Химический анализ отожженных пленок CdS, позволил идентифицировать монокристаллическое вещество с параметрами $a = 3{,}3$ Å; $c = 5{,}25$ Å как полиморфную гексагональную модификацию окиси кадмия, ранее в литературе не описанную.

Эпитаксиальные слои CdO гекс. и CdO_{NaCl} наблюдались также и на монокристаллах CdS после их отжига на воздухе.

Для выяснения механизма роста окисных соединений изучалось влияние ряда факторов на структуру, состав и электрофизические свойства (удельное сопротивление — ϱ и подвижность носителей заряда — μ) исходных и отожженных монокристаллических пленок и монокристаллов $A^{II}B^{VI}$.

Electrophysical properties of single crystals and films of $A^{II}B^{VI}$ compounds are determined not only by the conditions of their synthesis but also by the subsequent treatment, e. g. by annealing in air under a layer of activating powders that may result in structural and phase changes.

The number of papers concerned with the study of the structure and composition of single crystals and polycrystalline CdS films treated as mentioned above is limited.

As to similar investigations of single crystalline $A^{II}B^{VI}$ films they have not been reported at all.

The results obtained by some investigators by X-ray examination of polycrystalline films of cadmium sulphide annealed in the presence of oxygen under conditions that seemed to be the same are contradictory (BEHRINGER, CORSIN; SAKAI, OKIMURA; KAHLE, BERGER).

Cadmium sulphide is known to exist in three crystalline modifications: wurtzite, sphalerite and the cubic structure of the rock salt type.

As a rule the latter occurs under high pressure (ROOYMANS). There are no data as regards the growth of CdS_{NaCl} films. In this paper the study on changes in structure and composition of epitaxial $A^{II}B^{VI}$ films of various modifications is carried out by electron diffraction method. The changes occuring under heat treatment in air depend on a) condensation conditions; b) annealing conditions; c) the method of preparation and composition of initial CdS and CdSe powders.

For the sake of comparison a number of experiments with single crystals sawed along the plane and polished both mechanically and chemically (MAEDA, DOHEN, FUKUI) were performed.

1. Experimental

The starting materials for obtaining epitaxial films were a) CdS powders "for luminophores" synthesized by the wet method, i. e. by passing hydrogen sulphide through cadmium sulphate solution from which impurities of heavy metals had previously been removed $CdS^{(1)}$ and also the CdSe powder "for semiconductors" — $CdSe^{(1)}$ and CdTe(pure) — $CdTe^{(1)}$; b) $CdS^{(2)}$, $CdSe^{(2)}$ prepared from the vapour phase.

$A^{II}B^{VI}$ single crystalline films were grown by deposition from vapour on ordering substrates — either on the mica plane (001) (SERGEYEVA, KALINKIN, ALESKOVSKY 1967) or on the NaCl(111) plane — with the structure of wurtzite (CdS_W, $CdSe_W$), sphalerite (CdS_s $CdSe_s$ $CdTe_s$) or that of rock salt (CdS_{NaCl}) · CdS_{NaCl} single crystalline films grown on mica from $CdS^{(1)}$ powder by epitaxial method had orange colour and were in two twinning orientations turned through 180° around the direction [111] (Fig. 1). The CdS_{NaCl} plane (111) is parallel to the mica plane (001).

The lattice parameter $a = 5.6$ Å is close to parameter $a = 5.56$ Å — predicted by KABALKINA and TROIZKAYA for the CdS_{NaCl} films being under atmospheric pressure.

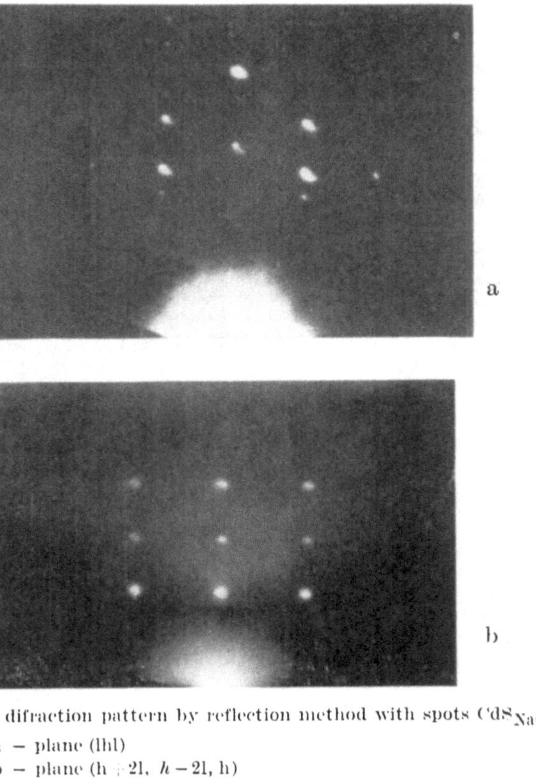

a

b

Fig. 1. Electron difraction pattern by reflection method with spots CdS_{NaCl}

a — plane (lhl)

b — plane (h + 2l, h − 2l, h)

2. Results

Intensity of spots relation of CdS_{NaCl} is in agreement with the data from GALKIN, TROIZKAYA, IVANOV in which texturized CdS_{NaCl} films were obtained on glass.

The annealing of single crystalline CdS_{NaCl} films in a muffle at a temperature of 400 °C for one hour results in the transformation of a part of the crystal into a wurtzite modification.

A net of dotted CdS_{NaCl}, CdS_{W} spots and additional spots due to the oxidation of CdS film (Fig. 2a) are observed on the electron diffraction patterns.

CdS_{NaCl} films can be kept under ordinary conditions for a long time but after electron bombardment in the electronograph *EG* — 100 A (for 10—15 min and a voltage 40—80 kV) they change into wurtzite modification in 1—3 days. Electron bombardement does not affect annealed CdS_{NaCl} films appreciably.

Single crystalline CdS films of various modifications were kept in the muffle at temperatures of 300—550 °C for 0,3—12 hours either under a layer of $CdS^{(2)}$ powder obtained from the vapour phase, or under a layer of activated (CdS) : CuCl (KALINKIN, SERGEYEVA, ALESKOWSKY) powders or without any powders.

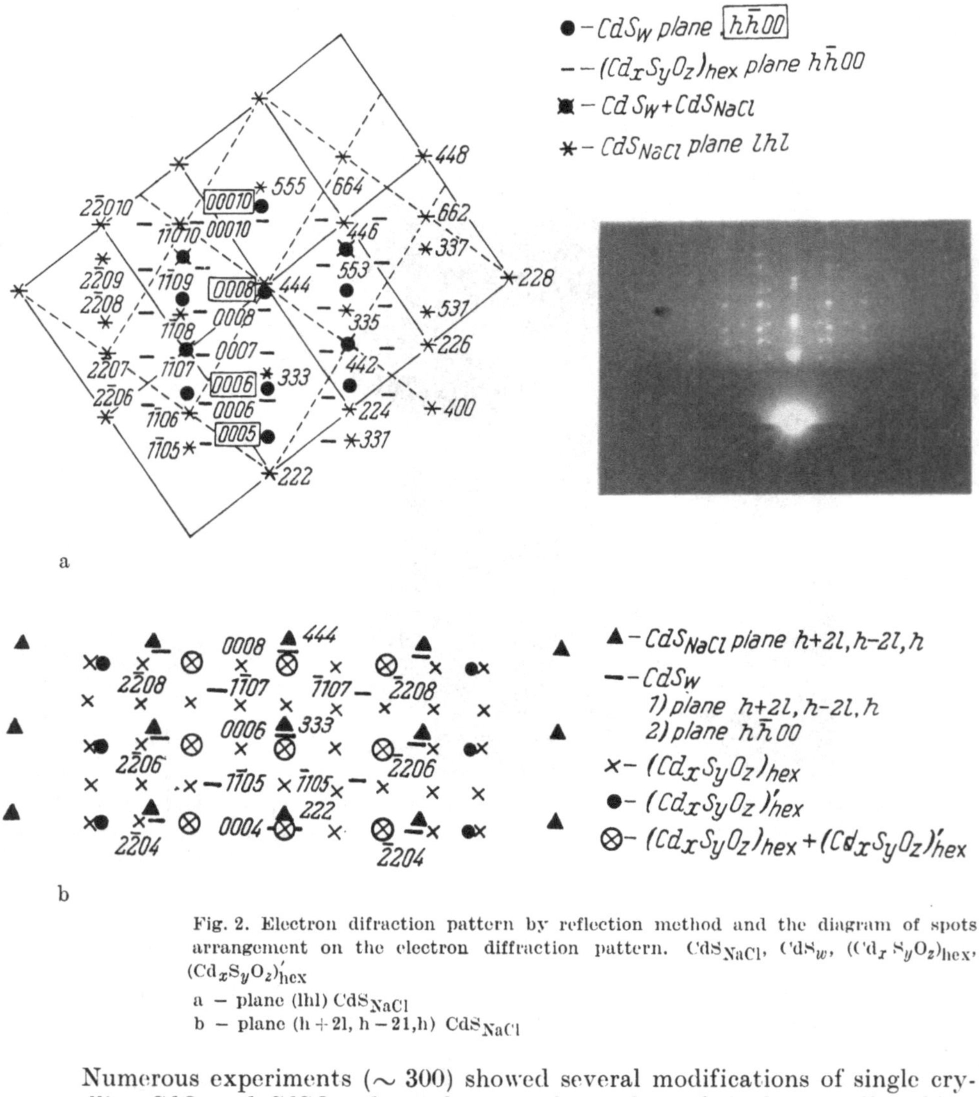

Fig. 2. Electron difraction pattern by reflection method and the diagram of spots arrangement on the electron diffraction pattern. CdS_{NaCl}, CdS_w, $(Cd_xS_yO_z)_{hex}$, $(Cd_xS_yO_z)'_{hex}$

a – plane (lhl) CdS_{NaCl}

b – plane (h + 2l, h − 2l, h) CdS_{NaCl}

Numerous experiments (~ 300) showed several modifications of single crystalline CdO and $CdSO_4$ salts to form on the surface of single crystalline films grown by sublimation of CdS[1] powder during annealing at temperatures 300–550 °C.

Oxidation of films begins at 320 °C; annealing at temperatures of 400–500 °C results in the formation of two single crystalline substances of hexagonal structure with close parameters $a = 5{,}3$ Å, $c = 7{,}0$ Å $(Cd_xS_yO_z)_{hex}$ and $a = 5{,}5$ Å, $c = 7{,}0$ Å $(Cd_xS_yO_z)'_{hex}$ on the surface of CdS films. In a number of cases weak additional spots are observed on the electron diffraction patterns from $(Cd_xS_yO_z)_{hex}$. These spots divide into two parts the distance between the center and the spots *h h* 2 *h l*. Their appearance seems to be connected with the doubling of the lattice parameter because of the lack of order in the periodicity of close packing layers, for example, of oxygen (Fig. 3a, 3b).

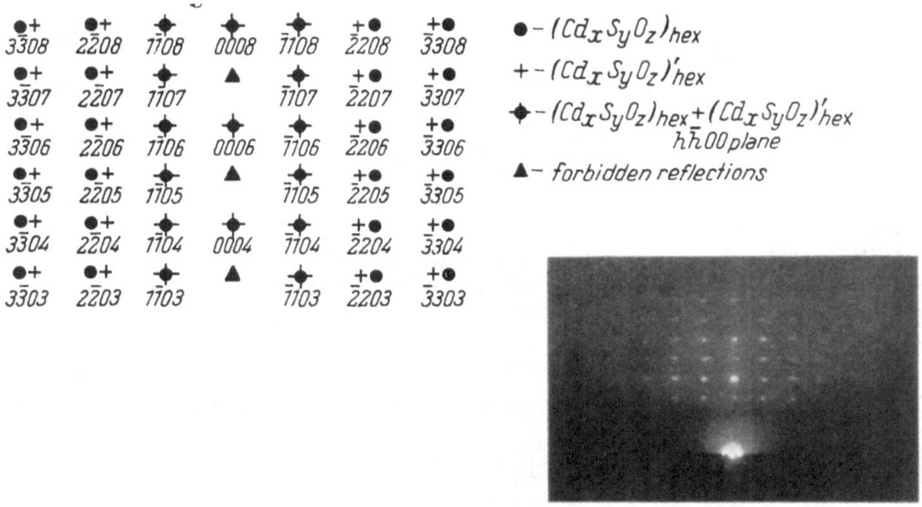

a

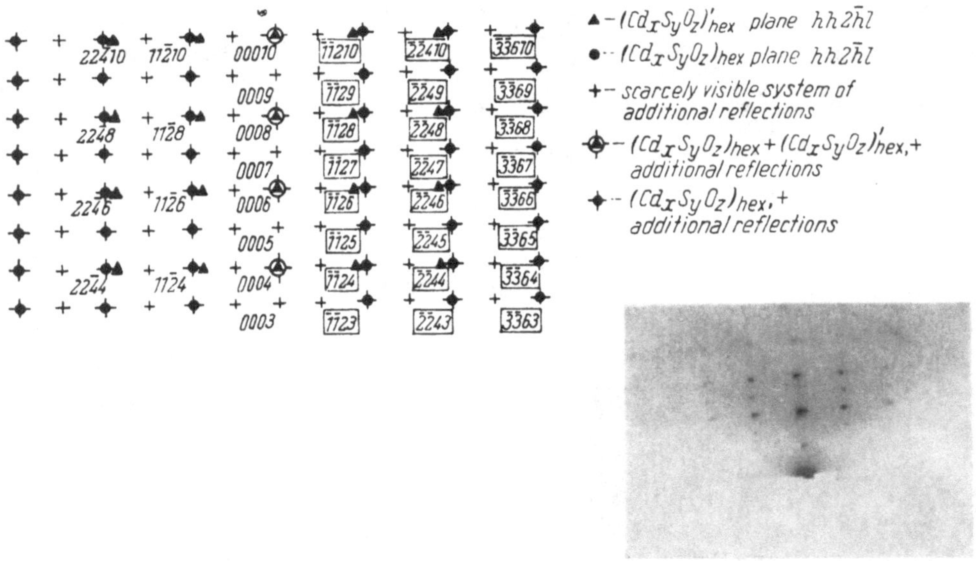

b

Fig. 3. Electron difraction pattern by reflection method and the diagram of spot arrangement on the electron difraction pattern $(Cd_xS_yO_z)_{hex}$, $(Cd_xS_yO_z)'_{hex}$
a – plane (hh00) $(Cd_xS_yO_z)_{hex}$
b – plane (hh2hl)

The spots on the electron diffraction pattern from $(Cd_xS_yO_z)$ hex and $(Cd_xS_yO_z)'_{hex}$ disappear if the film is washed with distilled water for 1–3 minutes. They seem to be due to cadmium sulphate ($CdSO_4$).

These data are in agreement with the results of BABINCHUK and SERDYUK who showed by electron microscope method that the crystalline aggregates with hexagonal or cubic forms appeared on the surface of CdS single crystals after heating in air at a temperature of 400°. They disappear after washing the crystal in distilled water. The above authors consider that on annealing the surface of the crystal is covered by a thin layer of $CdSO_4$. The layers similar to $(Cd_xS_yO_z)_{hex}$ were discovered before on annealed monocrystalline CdSe films (Sergeyeva, Kalinkin, Aleskowsky 1965). Within the same temperature range besides some oxycompounds $(Cd_xS_yO_z)_{hex}$, $(Cd_xS_yO_z)'_{hex}$ and others there grew ordinary cadmium oxide CdO'_{NaCl} in the form of texturized layers with the texture axis [001].

Within the temperature range investigated (300—550 °C) the annealing of the single crystalline CdS films is followed by the formation of single-crystalline substance of hexagonal structure (CdO_{hex}) with parameters $a = 3{,}3$ Å and $c = 5{,}25$ Å.

The chemical analysis of these films enabled this single-crystalline substance to be identified as an unknown polymorphic hexagonal cadmium oxide modification (Figs. 4a, 4b, 5).

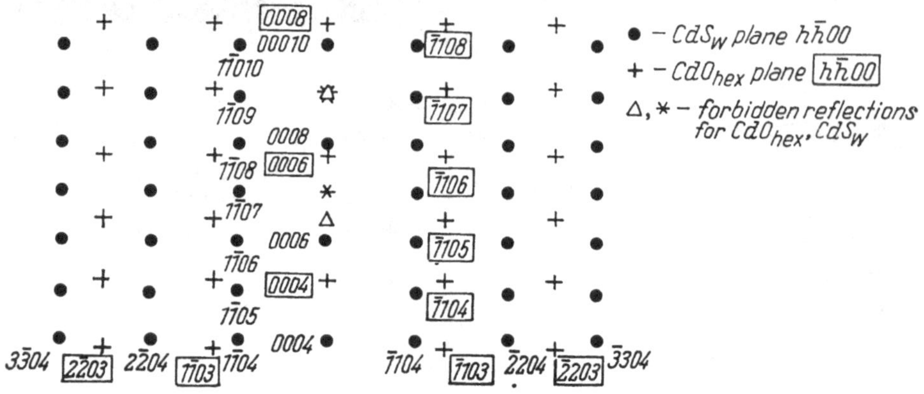

a

b

Fig. 4. Electron difraction pattern by reflection method and the diagram of spot arrangement on the electron difraction pattern CdO_{hex} and CdS_w
a – plane $(h\bar{h}00)$ CdO_{hex}, CdS_w
b – plane $(hh\overline{2h}l)$

At temperatures less than 450 °C CdO_{hex} crystallizes in small amounts together with other oxycompounds while at temperatures of 450—550 °C mainly this compound is observed on the electron diffraction patterns from CdS films.

The formation of single crystalline films consisting only of CdO_{NaCl} was observed comparatively rarely at temperatures less than 500 °C. CdO_{NaCl} crystallizes in the form of two twinning orientations turned through 180° round the direction [111]. CdO_{NaCl} plane (111) is parallel to CdS_w plane (0001), the

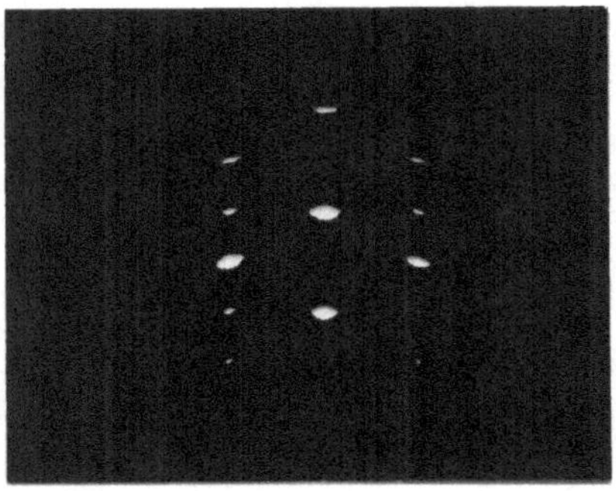

Fig. 5. Electron diffraction pattern by reflection method with spots CdO_{hex}, plane (hh00)

direction CdO_{NaCl} [1$\bar{1}$0] is parallel either to the direction CdS_w [11$\bar{2}$0] or to CdS_w direction [1$\bar{1}$00] (Figs. 6a, 6b, 6c, 6d).

When using $CdS_{(1)}$ powder of another batch the formation of CdO_{hex} single crystalline films was observed at annealing temperatures of 500 and 550 °C, the time of anannealing being 5—15 minutes. If annealing is continued for a longer time (20—80 minutes) texturized cadmium oxide CdO_{NaCl} begins to grow. The chemical composition of CdS films containing CdO_{hex} is: CdO — 3.08%, CdS — 97.11% at the annealing temperature 500 °C during 10 minutes and CdO — 3.15%, CdS — 96.77% at the annealing temperature of 550 °C during 5 minutes.

After prolonged washing (up to 27 hours) of annealed CdS films (containing either CdO_{hex} or CdO_{NaCl}) in distilled water both CdO_{hex} and CdO_{NaCl} spots could be seen on electron diffraction patterns and only after the film had been kept in a buffer solution $NH_4Cl + NH_4OH$ (pH = 8—10) for 1—3 minutes the spots disappeared and only dotted spots from CdS could be observed on the electron diffraction pattern.

Thus the hexagonal modifiation of cadmium oxide in solubility behaves as ordinary cadmium oxide CdO_{NaCl}. It should be noted that when annealing CdS films under batch (CdS): Cu, Cl the formation of cadmium oxide layers (both CdO_{NaCl} and CdO_{NaCl}) was not observed.

The following experiments were made to find out factors that influence the oxidation process of $A^{II}B^{VI}$ single crystalline films:

1. The influence of condensation conditions of $A^{II}B^{VI}$ films, the composition and electrophysical properties of initial and annealed films were studied.
2. The effect both of the method of preparation of original $A^{II}B^{VI}$ powders and their composition on the structure, composition and electrophysical properties of the films before and after annealed were studied.

The condensation conditions of cadmium sulphide films were altered as follows: a) constant temperature of the substrate being 270°, the temperature of the vaporizer was changed within the range of 800, 900, 1000 °C. b) constant temperature of the vaporizer being 1000 °C, the substrate temperature was 250, 300, 350, 400 and 450 °C.

Epitaxial CdS films prepared from the $CdS^{(2)}$ powder and condensed under the above conditions had wurtzite structure, and only when the substrate temperature was 270—300° the formation of films of mixed structure was observed (consisting of both wurtzite and sphalerite modifications) along with the formation of CdS_w films. As the substrate temperature increased the degree of perfection of CdS film structure improved. CdS films prepared from the $CdS^{(1)}$

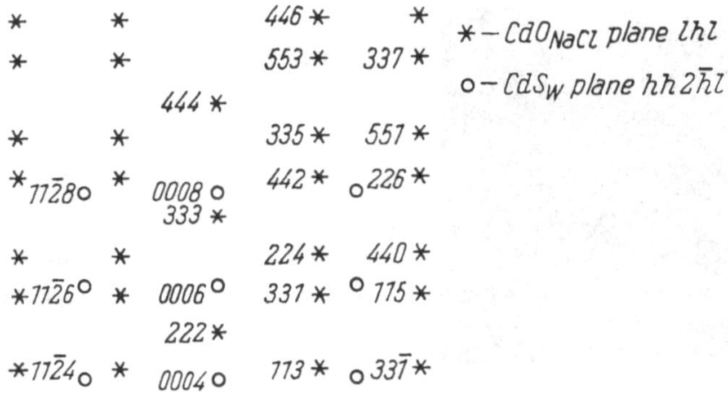

a

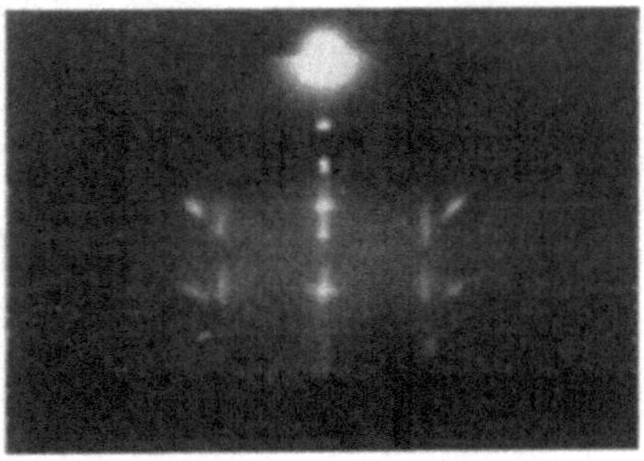

b

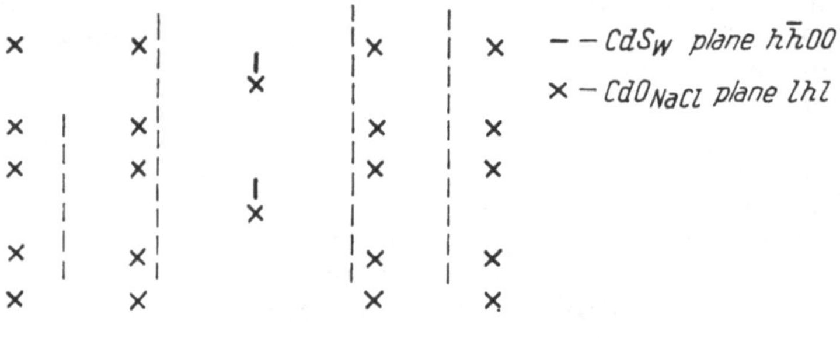

c

d

Fig. 6. Electron diffraction pattern by reflection method and the diagram of spot arrangement on the electron diffraction pattern CdO_{NaCl}, CdS_{w}.

a – planes (lhl) CdO_{NaCl}, (hh2hl) CdS_{w}

b – planes (h + 2l, h – 2l, h) Cd_{NaCl}, $(hh\overline{2h}l)$ CdS_{w}

c – planes (lhl) CdO_{NaCl}, (hh00) CdS_{w}

d – planes (hh00) CdO_{NaCl}, (hh2hl) CdO_{NaCl}

powder (The $CdS^{(1)\times}$ powder is of the same grade as that of $CdS^{(1)}$ but was obtained from another batch) under the above conditions contained sphalerite modification within the wider temperature range of the substrate, viz. $t_s^0 = 250-350$ °C. At temperatures of 300–350 °C films of sphalerite modification (CdS_S) were mainly formed. (Annealing of CdS films was carried out at temperatures of 450, 500, 550 °C during 10 to 120 minutes in a muffle). The decrease of supersaturation extent on condensing $CdS^{(2)}$ films results in the shortening of annealing temperature interval during which oxidation occurs; this may be due to the improvement of the degree of perfection of CdS film structure. As the substrate temperature increases the degree of perfection of the structure of now formed oxycompounds improves while the amount of oxidation products different in composition and orientation diminishes. Epitaxial growth of CdO_{NaCl} occurs mainly at an annealing temperature of 500–550 °C. Thus, if $CdS^{(2)}$ is condensed at a substrate temperature of 250 °C and at a vaporizing temperature of 1000 °C, oxidation occurs at an annealing temperatures of 450, 500, 550 °C; but at a substrate temperature of 450 °C oxidation followed by formation of CdO_{NaCl} takes place only at a temperature of 550 °C and during 10 min annealing. Oxidation of $CdS^{(1)}$ films begins under milder annealing conditions (450 °C, 30 min) as compared with those of $CdS^{(2)}$ film oxidation (450 °C, 120 min) and proceeds more intensively.

It has been shown that on condensing under similar conditions (the CdS powder prepared by different methods) the $CdS^{(1)}$ films are oxidized mainly both to cadmium sulphate (including a new polymorphous hexagonal modification) and basic sulphate whereas the $CdS^{(2)}$ films more frequently oxidized to ordinary cadmium oxide (CdO_{NaCl}, CdO'_{NaCl}).

Oxidation of epitaxial CdS films prepared from CdS powders—heat treated either in sulphur or cadmium vapour -did not show any marked variations. The study of electrophysical properties showed the specific resistance of CdS original films to change by 2–4 orders depending on both the way of preparing the initial CdS powder and the design of the ampule. ($CdS^{(2)} = 0.1-10$ ohm. cm $CdS^{(1)} = 10^2 - 10^3$ ohm. cm for the one ampoule design and $CdS^{(1)} = 10^5 - 10^7$ ohm. cm for the other ampoule design).

After annealing CdS films the specific resistance ζ was found to increase (for the films with a low value of ζ by several orders, for those having a high specific resistance only to a small extent) and became equal 10^4-10^7 ohm. cm. The mobility of charge carriers μ of original and annealed films was $10^{-2} - 50$ cm²/v · s, The thickness of the epitaxial films under investigation was 0.1–0.2 μm. Experiments carried out with CdS films of various thicknesses showed that within a thickness range of 0.1–1 μm there were no great differences as regards both specific resistance and charge carrier mobility of films as well as the formation of oxycompounds on their surface in the process of annealing.

Similar experiments were performed on CdSe and CdTe powders. Where condensation conditions for $CdSe^{(2)}$ were changed (substrate temperature varied from 200 to 550 °C, the vaporizing temperature from 500 to 450 °C) epitaxial films of CdSe grew having different degrees of perfection of structure. At the substrate temperature $t_s^0 = 400-550$ °C they are a wurtzite modification and at lower substrate temperatures t_s^0 400° a sphalerite one. The thickness of epitaxial films is 0.1–2.5 μm at an annealing temperature $t^0 = 420-550$ °C and during 5–120 min annealing they mainly oxidize to cadmium oxide CdO_{NaCl}. The oxidized CdO_{NaCl} layers are to be seen on electron diffraction patterns as

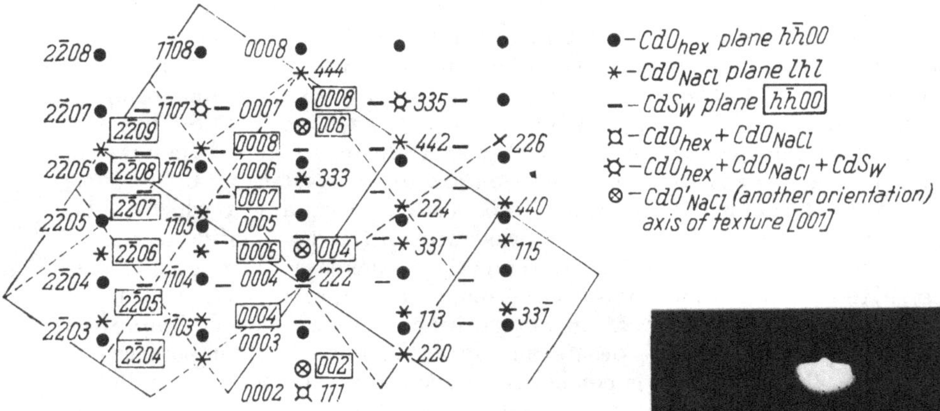

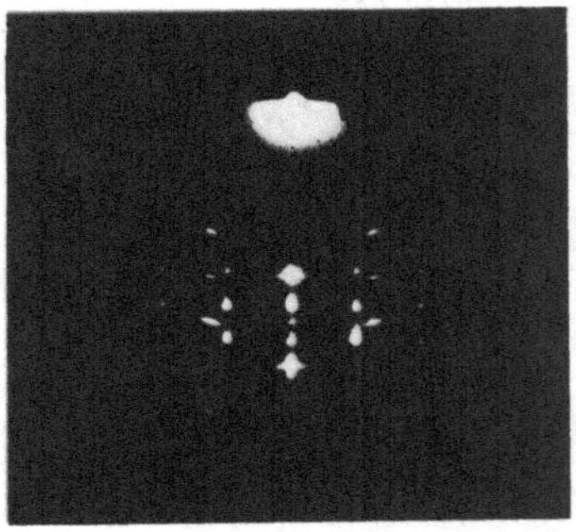

a

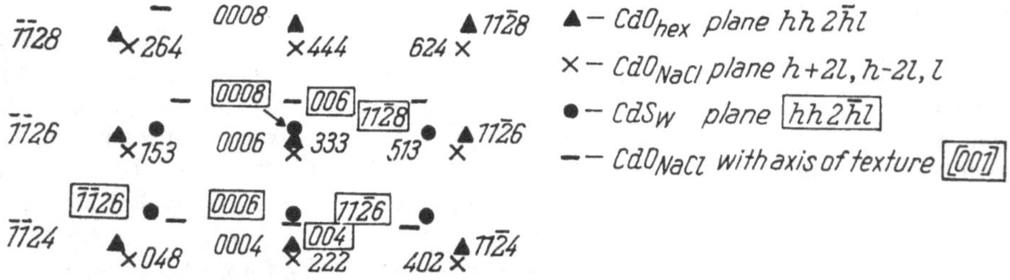

b

Fig. 7. Electron diffraction pattern by reflection method and the diagram of spot arrangement on the electron diffraction pattern CdO_{hex}, CdO_{NaCl}, CdS_w

a – planes $(h\bar{h}00)$ CdS_w, CdO_{hex}, (lhl) CdO_{NaCl}

b – planes $(hh2\bar{h}l)$, CdS_w, (h2l,h – 2l,h) CdO_{NaCl}

distinct polycristalline rings or as textures whose degree of perfection improves when that of CdSe epitax ialfilm structure also improves simultaneously with the increase of film thickness.

Epitaxial growth of CdO_{NaCl} is seldom observed at annealing temperatures of 550 – 600 °C.

CdSe films prepared from CdSe[1] powder are likely to form upon oxidation textures corresponding to cadmium salt $CdSeO_3$. The annealing of single crystalline CdTe[1] films at temperatures of 450 – 600 °C gives rise to formation of texturized layers of cadmium oxide with texture axis [001] (CdO'_{NaCl}), single crystalline layers CdO_{NaCl} as well as to formation of textures of a more complex composition ($Cd_xTe_yO_z$).

3. Summary

1. Single crystalline CdS_{NaCl} films have been grown on the mica plane (001) by epitaxial method.

2. The formation of oxygen compounds on single crystalline $A^{II}B^{VI}$ film surface as well as the composition of the above compounds depend on the method of synthesis, the composition and structure of original cadmium powder.

3. Decrease of supersaturation with the condensation of epitaxial CdS films prepared from $CdS^{(2)}$ powder results in both improvement of degree of perfection of their structure and shortening of the temperature range at which CdS film oxidation accurs.

4. Oxidation of films and $A^{II}B^{VI}$ crystals took place at much lower temperatures than that indicated in literature.

5. Epitaxial single crystalline films of hexagonal oxide and cadmium sulphate of polymorphous modifications which were unknown before grew on the surface of single crystalline CdS films. The $CdS^{(1)}$ powder synthesized by the "wet" method has been used for the sublimation of the above films. In a number of cases films have also been grown on single crystals by annealing in air.

6. Annealing of single crystalline CdS films and CdSe of cubic modification simultaneously with their oxidation was followed by the change of their structure to the hexagonal one.

7. Growth of oxyfilms was not affected appreciably by the modification of the original CdS films. For example, at a temperature of 450—550° and annealing period from 0.3 to 12 hours CdS_S films have changed completely or partially to CdS_W, the spots of which on electron diffraction patterns were accompanied by those of the same oxycompounds as in the case of the use of CdS_W films from the start.

8. Epitaxial films of cadmium oxide have been obtained during the annealing period in air of CdS, CdSe, CdTe single crystalline films.

Authors express their sincere gratitude to professor L. S. Palatnik for discussing the results of this paper.

Students N. I. Fershter and T. P. Kazannikova tock part in these investigations.

References

Babinchuk, V. S., Serdyuk, V. V.: FTT **10**, (2), 645 (1968)

Behringer, A. J., Corsin, L.: J. electrochem. Soc. **110**, 1038 (1963)

Galkin, B. D., Troizkaya, N. V., Ivanov, R. D.: Kristallografiya **12**, (5), 878 (1967)

Kabalkina, S. S., Troizkaya, E. V.: Doklady AN SSR **151**, 1068 (1963)

Kahle, W., Berger, H.: Physica Status solidi **14**, K 201—K 204 (1966)

Kalinkin, I. P., Sergeyeva, L. A., Aleskovsky, V. B.: Izvestia Akad. Nauk UdSSR Neorganicheskie material **2**, 2110 (1966)

Maeda, M., Dohen, M., Fukui, H.: Japan. J. appl. Physics **6**, 652 (1967)

Rooymans, C. J. M.: Philips Res. Rep. Suppl.: Nr. 5, 1—135 (1968)

Sakai, Y., Okimura, H.: Japan. J. appl. Physics **3**, 144 (1964)

SERGEYEVA, L. A., KALINKIN, I. P., ALESKOVSKY, V. B.: Kristallografiya **10**, 237 (1965)
SERGEYEVA, L. A., KALINKIN, I. P., ALESKOVSKY, V. B.: Kristallografiya **12**, (1), 113 (1967)

(Received April 28, 1969)

Authors' address:

Dr. L. A. SERGEYEWA
Dr. I. P. KALINKIN
Prof. Dr. W. P. ALESKOWSKI
A. P. NECHIPORENKO
Institute of Technology (Lensovieta)
Leningrad L-13, Sagorodnyj Prospect 49
USSR

Kristall und Technik	5	1	1970	73–84

G. I. Distler

Institute of Crystallography, Academy of Sciences of the U.S.S.R., Moscow

Electrical Structure of the Surface of Crystalline Substrates and its Influence on Nucleation and Growth Processes

The electrical structure of crystalline substrates has been studied by a new technique based on selective crystallization of decorating material and selective deposition of charged colloidal particles. Silver chloride was found to be a very effective indicator of the electrical relief of crystal surfaces. At early stages of thermal condensation, it crystallizes orientedly on negatively charged regions of the surface and randomly on those charged positively. The method has been applied to the study of triglycine sulphate, LiF, and NaCl crystals. It has been found that the surface of real crystals consists as a rule of positive and negative regions, representing assemblies of charged point defects. Near the cleavage steps certain "dead zones" arise, in which a local compensation of charges takes place. Nucleation occurs mainly at charged point defects and oriented coalescence of islands proceeds with different velocity on surface local regions of different signs. The selectivity of crystallization depends on the relation of the sign and magnitude of potential between different smooth areas of the surfaces of crystalline substrates, on the one hand, and between steps and smooth areas, on the other. The charged point defects and their assemblies which are present on the surface of crystalline substrates induce polarization structures of the electret type in amorphous interfacial layers. These structures are "frozen" in interfacial layers in so stable a manner that the layers turn out to be electrical copies of crystal surfaces, thus reflecting the distribution of potential of the surface. Such copies preserve the "electrical memory" even after they are detached from the surfaces of crystals and can hence be used as active substrates for various heterogeneous processes, in particular, for epitaxy.

Разработаны новые методы изучения электрической структуры кристаллов-подложек, основанные на избирательной кристаллизации декорирующих веществ, а также на избирательном осаждении заряженных коллоидных частиц. Весьма эффективным индикатором на электрический рельеф поверхности кристаллов оказалось хлористое серебро, которое на ранних стадиях термической конденсации кристаллизуется ориентированно на отрицательно заряженных участках поверхности и беспорядочно на положительных участках. Изучен электрический рельеф поверхности кристаллов триглицинсульфата, LiF и NaCl. Установлено, что поверхность реальных кристаллов, как правило, состоит из положительных и отрицательных участков, представляющих собой скопления заряженных точечных дефектов. Вблизи ступеней скола возникают ,,мертвые зоны", в которых имеет место локальная компенсация зарядов. Зародышеобразование происходит преимущественно на заряженных точечных дефектах, а ориентированная коалесценция зародышевых кристаллов протекает с различной скоростью на локальных участках поверхности разного знака. Избирательность кристаллизации определяется соотношением знака и величны потенциала между различными гладкими участками поверхности кристаллов-подложек, с одной стороны, и ступенями и гладкими участками, с другой стороны. Заряженные точечные дефекты и их скопления на поверхности кристаллов-подложек индуцируют в аморфных граничных слоях поляризационные структуры электретного типа. Эти структуры „замораживаются" в граничных слоях настолько устойчиво, что такие слои превращаются в электрические копии поверхности кристаллов, отражающие распределение потенциала поверхности. Такие копии сохраняют „электрическую память", даже будучи отделенными от поверхности кристаллов и могут быть использованы в качестве активных подложек для различных гетерогенных процессов, в частности, для эпитаксии.

Until recently the nucleation and growth processes in crystals and thin films have been generally considered from the point of view of the microgeometry of the surface (cleavage and growth steps, emergent sites of screw dislocations). The point defects commonly have not been taken into consideration, the same as the electrical properties of the surfaces of crystalline substrates. No attempts have been made to relate the electrical relief of the surface of crystals to its geometrical relief and to study the separated and combined influence of these reliefs on nucleation and growth processes.

We earlier found (Distler (1); Distler, Zvyagin; Distler (2)) that electrically active point defects play the determining part in nucleation. Fairly convincing proof for this has been obtained from decorating the surface of X-Ray-irradiated alkali halide crystals (Distler (3); Distler, Lebedeva, Moskvin), when nucleation occurs on color centres, and from decorating the surface of a ferroelectric — triglycine sulphate — above the Curie point (Distler, Konstantinova, Gerasimov, Tolmacheva (1), (2)), when the "memory" of the domain structure, which is a totality of regularly positioned point defects, is preserved.

In this laboratory new methods of detecting the electrical structure of crystalline substrate surfaces have been developed. They are based on selective crystallization of decorating material on electrically different regions of the surface and also on selective interaction of charged colloidal particles with such regions. Silver chloride proved to be a very effective indicator of the electrical relief of the surface, for at early stages of thermal condensation it crystallizes orientedly on negatively charged regions of the surface and randomly on those which are charged positively (Distler, Konstantinova, Vlasov; Distler, Vlasov (1)). This offers a possibility of detecting the electrical relief not only by analysis of decoration pictures but also by diffraction.

As model crystals having a known electrical structure we used crystals of a ferroelectric — triglycine sulphate (TGS), the domain structure of which can be determined by independent methods (Konstantinova). The evaporation rate of AgCl was 8 Å s^{-1}; during evaporation the TGS crystals were maintained at room temperature. At the earliest condensation stages (up to a mean layer thickness $\sim$ 10 Å), discrete AgCl nuclei originate, positioned randomly with respect to the surface of the domains of both signs. The density of the nuclei is $5 \cdot 10^{10}$ cm^{-2}, this being consistent with the density of point defects in ionic crystals. From the polarity of TGS crystals in the direction [010] which is perpendicular to cleavage surfaces follows the polarity of active centres which under given conditions do not exhibit orienting influence. With an increase in the mean layer thickness, oriented coalescence of AgCl nuclei occurs differently on domains of different signs (Fig. 1). Oriented coalescence of nuclei starts on

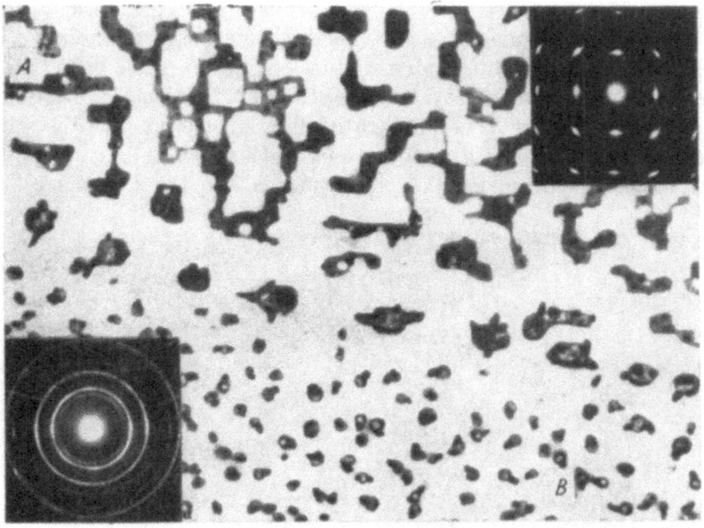

Fig. 1. Decoration of a triglycine sulphate surface with silver chloride. Oriented coalescence of nuclei on the negative domain (A); no orientation of nuclei on the positive domain (B). Mean AgCl layer thickness 40 Å

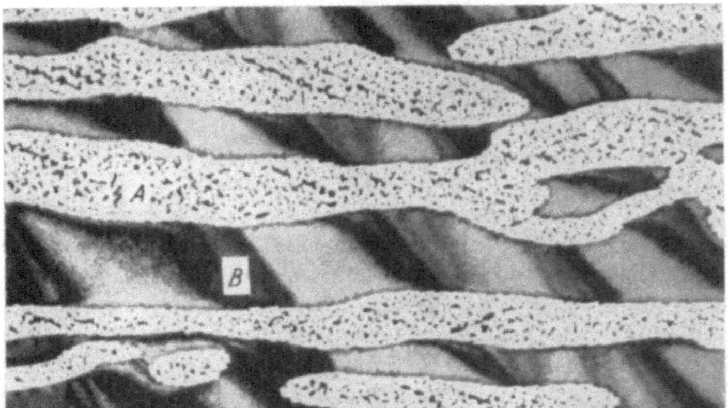

A ~ B

Fig. 2. Formation of monocrystalline silver chloride film on the surface of the negative domain of TGS (A). No preferential AgCl orientation on the surface of the positive domain (B). Mean AgCl layer thickness 90 Å

the surface of negative domains: the particles of AgCl join together and, besides, narrow bridges form between them, described by several authors (Poppa; Pashley) (Fig. 1). On the surface of positive domains up to a mean AgCl layer thickness of 90 Å only an enlargement of AgCl nuclei takes place, with no preferential orientation (Fig. 2). At this AgCl layer thickness, on the surface of negative domains a continuous monocrystalline AgCl film already develops (Fig. 2), while on the surface of positive domains the formation of a monocrystalline film begins at a mean layer thickness of 150 Å. The electrical relief of the surface influences crystallization of AgCl not only on flat regions but also on cleavage steps. Figure 3 demonstrates that on cleavage steps which intersect the surface of domains of different signs AgCl deposits in a larger or in a lesser amount than on flat regions of the domains. In the first case the step looks

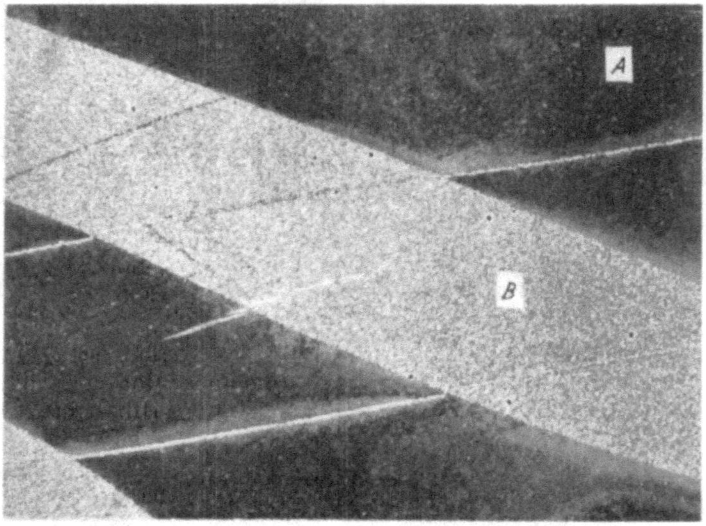

Fig. 3. Selective crystallization of AgCl on charged cleavage steps of TGS surface. (A) – negative domain; (B) – positive domain. Mean AgCl layer thickness 90 Å

darker and in the second case it looks lighter than the flat regions. The selectivity of deposition on flat regions is associated with oriented coalescence of AgCl which occurs primarily on negative domains. The different deposition rate of AgCl on different cleavage steps may be, by analogy, explained by that the sign and magnitude of the step charge can coincide with or differ from corresponding electrial characteristics of flat regions of the surface. Crystallization of AgCl occurs mostly on those areas of the surface — be they steps or flat regions — which have the largest negative potential. Crystallization pictures show that in many cases precisely the electrical structure of the surface of crystalline substrates, and not its microgeometry, accounts for nucleation and growth processes. The determinant factor in oriented crystallization is the relation of the sign and magnitude of potential between electrically different flat regions of the surface, on the one hand, and between steps and flat regions, on the other. Therefore crystallization at early stages primarily can proceed selectively either only on steps, or only on flat regions, or, if the electrical properties of steps and flat regions are identical, simultaneously on both elements of the surface.

The developed method of visualization of the electrical relief of the surface has been applied to other ionic and semiconductor crystals. In particular, single crystals of LiF (DISTLER, VLASOV) and NaCl (DISTLER, SAROVSKY) have been studied in detail. Selective crystallization of AgCl on cleavage surfaces of these single crystals results in detection of areas differing by morphology and orientation of decorating crystals. At initial stages of deposition, oriented coalescence of AgCl occurs differently on local regions of the surface (Figs. 4 and 5), just as in the case of deposition on the surface of TGS. On these regions, the potential of which may be considered (by analogy with TGS) as the most negative, sufficiently large AgCl crystals arise, around of which empty regions, so-called "dead zones" are formed (Fig. 4). On the other (positive) regions of the

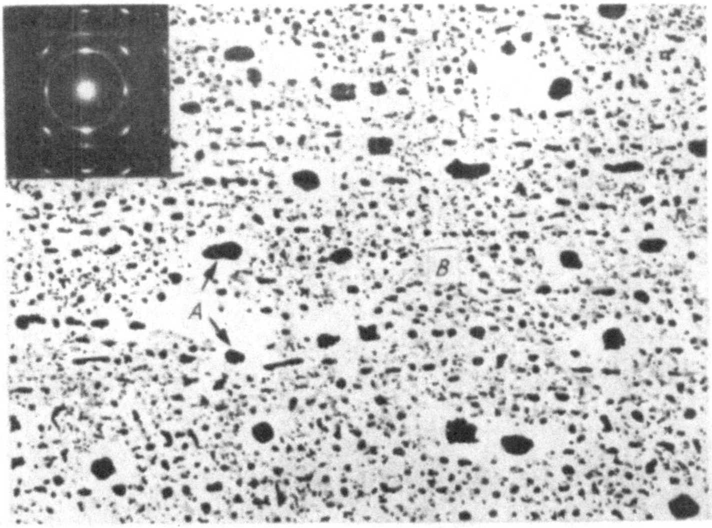

Fig. 4. Electrical relief of LiF cleavage surface decorated with AgCl. (A) – negative region; (B) – positive region of the surface. Mean AgCl layer thickness 30 Å

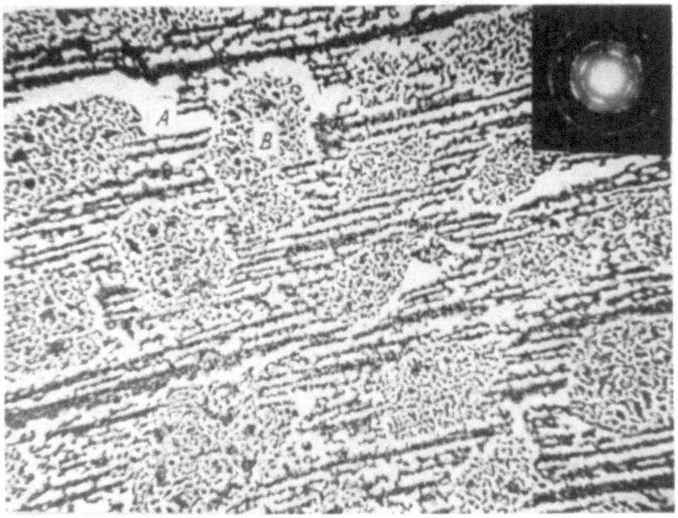

Fig. 5. Electrical relief of NaCl cleavage surface decorated with AgCl. (A) – On negative regions of NaCl surface the crystals of AgCl are positioned orientedly. (B) – On positive regions the crystals of AgCl are positioned randomly. Mean AgCl layer thickness 60 Å

surface at this stage smaller AgCl crystals arise, non-oriented relative to LiF substrate surface. Charged regions of different signs on LiF surface are of the size of several thousands of Angströms (Fig. 4), whereas on NaCl surface they reach the size of several microns (Fig. 5). Analysis of our results (DISTLER, GERASIMOV) and those from literature (see, for example, (GRITSAENKO, ZVYAGIN, BOYARSKAYA, GORSHKOV, SAMOTOIN, FROLOVA; ADAMSKY) indicates that oriented crystallization not only of silver chloride but also of many other extensively used substances, such as gold, silver, lead telluride, lead sulphide and others proceeds with different velocity and by different mechanisms on local regions which differ from one another in sign and magnitude of potential.

We note a very frequent case when in the vicinity of cleavage steps no crystallization of decorating material occurs, "dead zones" develop. Analogous zones are also observed at electrical decoration of the surface of TGS crystals

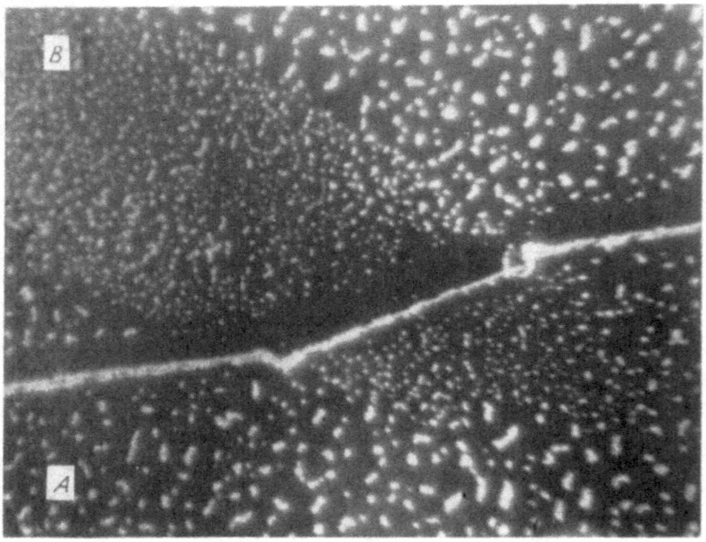

Fig. 6. Selective deposition of negatively charged polystyrol latex of the size 7500 Å on a TGS cleavage surface. (A) – negative domain; (B) – positive domain

with charged particles of polystyrol latex from water-acetone solutions. In this case the latex particles deposit on the surface of the domains of both signs (DISTLER, TOKMAKOVA) (Fig. 6). This indicates that the "dead zones" differ in their electrical properties from positive and negative regions of the TGS surface. "Dead zones" are also observed at decoration of NaCl and LiF crystal surfaces with charged latex (Fig. 7). It seems that the "dead zones" are neutral regions of crystal surfaces in which a local compensation of charges takes place. These zones, naturally, possess the least reactioning ability: the atoms and molecules of depositing substances do not stay on them but migrate to most active charged regions of the surface. The charged regions of the surface are assemblies of charged vacancies, interstitial atoms and impurities: to provide the neutrality of the crystal as a whole, they are compensated by defects of opposite sign over fairly large distances, as considered in the paper of LIFSHITS, GEGUZIN; KRIVOGLAZ and evidenced experimentally by DISTLER, VLASOV (2); DISTLER, SAROVSKY; KORNFELD.

Fig. 7. Selective deposition of negatively charged polystyrol latex of the size 900 Å on a NaCl cleavage surface

The electrical properties of crystal surfaces are also closely associated with the long range mechanism of crystallization, consisting in orienting effect of the real surface of crystal substrates through diffractionally amorphous interfacial layers. Crystallization by the long range mechanism has been observed in thermal evaporation (GERASIMOV, DISTLER; DISTLER, KOBZAREVA, GERASIMOV; KHIDR, IGNACZ, POCZA), cathodic sputtering (GERASIMOV, DISTLER, EFREMENKOVA, YURASOVA) and deposition from solutions (DISTLER, KOBZAREVA; DISTLER, FEDOTOVA). It has not been established, however, whether amorphous interfacial layers act merely as an inert intermediate medium or whether they themselves provide an informational net.

To ascertain the mechanism of the long range effects we performed a series of experiments. Intermediate interfacial layers were prepared from material with different electrical properties. Then, after separation from the surface of crystals on which they were formed, their electrical structure was investigated by means of decoration. In the first place, we studied such substances the electrical properties of which can provide or, conversely, do not provide the occurrence of thermoelectret state. Interfacial layers were prepared from diffractionally amorphous plastic films which exhibit (BINDER) (polyvinyl chloride — PVC) and do not exhibit (WISEMAN, LINDEN) (nitrocellulose — NC) thermoelectret properties. NaCl and mica single crystals served as substrates; as crystallizing material served antrachinon which crystallizes on these substrates as needleshaped crystals (WILLEMS; ROYER), thus enabling to establish visually the presence or the absence of orientation at optical scale. Polymer films of about 1000 Å thick were prepared on fresh cleavage surfaces of crystals, then detached from them in water, and on to the contact side of the films antrachinon was sublimated. Antrachinon was sublimated also directly on to the surface of crystals, under identical conditions.

Directly on the surface of NaCl the needle-shaped crystals of antrachinon form mainly biaxial (Fig. 8a) and on the surface of mica uniaxial (Fig. 8b) textures. [110] of NaCl is parallel to [001] of antrachinon, and [100], [110] of mica is parallel to [001] of antrachinon. On the contact side of non-heated PVC and NC films, as well as of pre-heated above the softening point NC films, antrachinon crystallizes randomly. Only on the contact side of pre-heated above the softening point PVC films biaxial and uniaxial antrachinon textures,

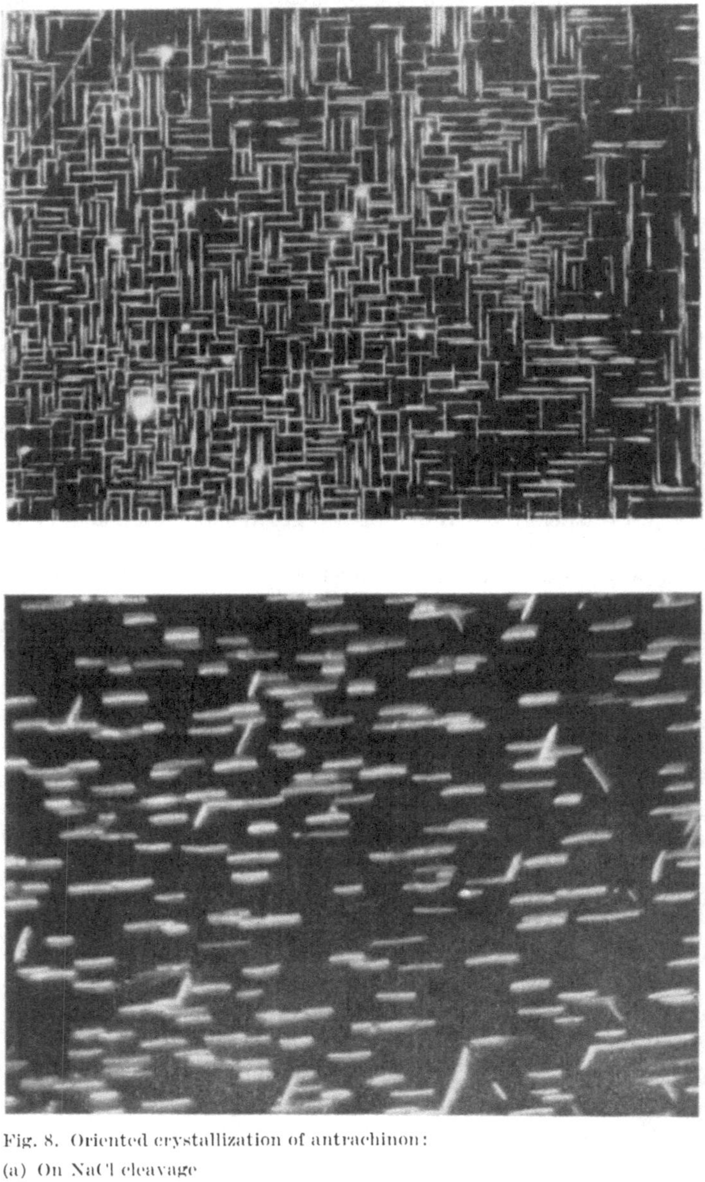

a

b

Fig. 8. Oriented crystallization of antrachinon:

(a) On NaCl cleavage

(b) On mica cleavage

respectively, are formed (Figs. 9a and 9b), identical to those obtained directly on the surface of NaCl and mica crystals. The occurrence of oriented crystallization of antrachinon solely on the contact side of PVC films which had been previously heated above the softening point (this being necessary for the occurrence of thermoelectret properties) proves that the mechanism of "remembering" the orientational information is due to electret properties of amorphous interfacial layers (DISTLER, KOBZAREVA (1), (2). The role of external electric

a

b

Fig. 9. Oriented crystallization of antrachinon:
(a) On the contact side of a pre-heated to ~ 110 °C polyvinyl chloride film, stripped from the surface of NaCl
(b) On the contact side of a pre-heated to ~ 110 °C polyvinyl chloride film, stripped from the surface of mica

field is played in this case by local electric micro- and macrofields of charged point defects and their assemblies. The tensions of these fields over atomic distances may reach hundreds and thousands of kV cm^{-1}. This is known to be sufficient for producing an induced polarization in interfacial layers. The occurrence of polarization structures is a kinetic process and therefore the conditions of preparation the interfacial layers essentially predetermine the character and the degree of "remembering" the electrical properties of the surface by these layers.

The *p–n*-junction on the surface of silicon single crystals (DISTLER, SHENYAVSKAYA (1), (2)) was also used as an active surface which forms the electrical structure of interfacial layers. Amorphous interfacial layers were obtained by evaporation of carbon from an arc. Crystallization of AgCl by thermal evaporation directly on the surface of silicon gives rise to a monocrystalline film (about 700 Å thick) on the *p*-region and to a thinner (about 100 Å thick)

Fig. 10. Decoration with silver chloride of the contact side of a carbon film, detached from the surface of silicon; the *p–n*-junction is reproduced

polycrystalline film on the *n*-region. If crystallization of AgCl is performed on the contact side of the carbon film detached from the silicon surface by means of gelatine, the geometry of the *p–n*-junction is seen to be reproduced very exactly, as well as the selectivity of crystallization of AgCl on those regions of carbon film which arose on *p*- and *n*-regions of silicon surface, respectively (Fig. 10). The observed effect of selective crystallization of AgCl on the contact side of amorphous carbon layers may also be explained by the occurrence in them of an internal polarization of the electret type. One may believe that the electrophysical properties of carbon in thin interfacial layers which develop on the surface of crystals under the influence of very strong electric fields of charged defects might essentially differ from the properties of carbon in macrospecimens.

So far crystallization by the long range mechanism has been effectuated through interfacial layers of very different nature (plastics, carbon, silicon monoxide, silicon dioxide and others), one may deduce that the occurrence in interfacial layers of an induced internal polarization which reflects the electrical microrelief of surfaces of crystalline substrates is a common phenomenon. This induced polarization is "frozen in", as experiments show, in so stable a manner that the interfacial layers turn out to be peculiar copies of crystal surfaces which retain the "electrical memory" even after separation from substrate surfaces. The comparison of the properties of solid interfacial layers with those of the liquid interfacial layers, studied in detail in many references (DERYAGIN (ed.) 1963, 1967) shows that a profound analogy exists between them. Both types of layers may be considered in terms of electrical structure of the surface of real solids. There exists apparently a close connection between these layers and the corresponding transitional semi-ordered layers occurring near the surface of a crystal growing in solution or melt (GYULAI, BIELEK; GYULAI). The totality of experimental data confirms the volumetric character of crystallization processes which manifests itself, in particular, in the possibility of crystallization by the long range mechanism without a direct contact of crystallizing material with the surface of crystal substrate to be necessary. The possibility of obtaining electrical copies of the surface of crystalline substrates on the base of the occurrence of induced polarization structures in diffractionally amorphous interfacial layers offers, in our opinion, encouraging prospects of using the interfacial layers as active surfaces in various heterogeneous processes, epitaxy included.

References

ADAMSKY, R. F.: The Use of Thin Films in Physical Investigations, ed. J. C. Anderson, London 1966, p. 243

BINDER, F.: Z. Naturforsch. **6a**, 714 (1951)

DERYAGIN, B. V.: (ed.) Research in Surface Forces, I, New York 1963; ibid., 2 (1966)

DERYAGIN, B. V. (ed.): Research in Surface Forces, Moscow 1967 (in russian)

DISTLER, G. I.: In Collection Growth of Crystals, vol. 8, Moscow 1968, p. 108 (in russian) (1)

DISTLER, G. I.: Izv. AN SSSR, Ser. Fiz. **32**, 1044, 1968 (2)

DISTLER, G. I.: Proc. Second Intern. Conf. Cryst. Growth, Birmingham 1968, published in J. Crystal Growth **3,4**, 175 (1968) (3)

DISTLER, G. I., FEDOTOVA, M. P.: Vysokomolek. soed., ser. B, No. 1, 3 (1967)

DISTLER, G. I., GERASIMOV, Y. M.: Kristallografiya **5**, 346 (1970)

DISTLER, G. I., KOBZAREVA, S. A.: Dokl. AN SSSR **172**, 1069 (1967)

DISTLER, G. I., KOBZAREVA, S. A.: Naturwissenschaften **56**, 325 (1969)

DISTLER, G. I., KOBZAREVA, S. A.: Dokl. AN SSSR **88**, 811 (1969) (2)

DISTLER, G. I., KOBZAREVA, S. A., GERASIMOV, Y. M.: J. Crystal Growth **2**, 45 (1968)

DISTLER, G. I., KONSTANTINOVA, V. P., GERASIMOV, Y. M., TOLMACHEVA, G. A.: Nature **218**, 762 (1968) (1)

DISTLER, G. I., KONSTANTINOVA, V. P., GERASIMOV, Y. M., TOLMACHEVA, G. A.: Izv. AN SSSR, Ser. Fiz. **32**, 1252 (1968) (2)

DISTLER, G. I., KONSTANTINOVA, V. P., VLASOV, V. P.: Kristallografiya **14**, 90 (1969)

DISTLER, G. I., LEBEDEVA, V. N., MOSKVIN, V. V.: Fiz. Tverd. Tela **10**, 3489 (1968)

DISTLER, G. I., SAROVSKY, E. G.: Fiz. Tverd. Tela **11**, 547 (1969)

DISTLER, G. I., SHENYAVSKAYA, L. A.: Nature **221**, 52 (1969) (1)

DISTLER, G. I., SHENYAVSKAYA, L. A.: Fiz. Tverd. Tela **11**, 488 (1969) (2)

DISTLER, G. I., TOKMAKOVA, E. I.: Kristallografiya **14**, 1055 (1969)

DISTLER, G. I., VLASOV, V. P.: Thin Solid Films **3**, 333 (1969) (1)

DISTLER, G. I., VLASOV, V. P.: Fiz. Tverd. Tela **11**, 2226 (1969)

DISTLER, G. I., ZVYAGIN, B. B.: Nature **212**, 807 (1966)

GERASIMOV, Y. M., DISTLER, G. I.: Naturwissenschaften **55**, 732 (1968)

GERASIMOV, Y. M., DISTLER, G. I., EFREMENKOVA, V. M., YURASOVA, V. E.: Fiz. Tverd. Tela **10**, 270 (1968)

GRITSAENKO, G. S., ZVYAGIN, B. B., BOYARSKAYA, R. V.. GORSHKOV, A. I., SAMOTOIN, N. D., FROLOVA, K. G.: Methods of Electron Microscopy of Minerals, Moscow 1969 (in russian)
GYULAI, Z., BIELEK, S.: Acta physica Acad. Hung. **1**, 199 (1952)
GYULAI, Z.: In Growth of Crystals, vol. 3, Moscow 1961, p. 98 (in russian)
KHIDR, M. S., IGNACZ, P., POCZA, J.: Proc. 2nd Colloqu. Thin Films, Budapest 1967, p. 56
KONSTANTINOVA, V. P.: Kristallografiya **7**, 748 (1962)
KORNFELD, M. I.: Fiz. Tverd. Tela **10**, 2422 (1968)
KRIVOGLAZ, M. A.: Fiz. Tverd. Tela **10**, 3348 (1968)
LIFSHITS, I. M., GEGUZIN, Y. E.: Fiz. Tverd. Tela **7**, 62 (1965)
PASHLEY, D. W.: Advances in Physics **14**, 327 (1965)
POPPA, H.: Z. Naturforsch. **A 19**, 835 (1964)
ROYER, L.: C. R. Acad. Sci. **236**, 1804 (1953)
WILLEMS, I.: Naturwissenschaften **31**, 146 (1943)
WISEMAN, G. G., LINDEN, E. G.: Electr. Engng. **72**, 869 (1953)

(Received April 28, 1969)

Authors address:

Prof. Dr. G. I. DISTLER
Institute of Crystallography
Academy of Sciences of the USSR
Moscow B-333
Lenin Prospect 59

| Kristall und Technik | 5 | 1 | 1970 | 85–93 |

J. Pfeifer, L. Gútai, É. Sz. Harraszthy

Research Institute for Technical Physics of the
Hungarian Academy of Sciences, Budapest

The Structure of Epitaxially Grown Layers and the Electrical Parameters of $p-n$ Junctions Formed on them

Germanium layers were evaporated onto p-type substrates in oil diffusion pumped systems under pressures varying between 3×10^{-5} and 1×10^{-8} torr. Evaporation parameters: CP-4 etched (111) Ge substrate; 810 °C substrate temperature; condensation rate: $3-5$ μmin^{-1}; layer thickness: $15-25$ μ. Structural properties were determined by etching rate, microscopy and electron diffraction examinations. Graded $p-n$ junctions were formed by diffusion into the layers. I–V, C–V characteristics of the junctions and storage-time were determined. Hole concentrations between 1×10^{18} and 4×10^{15} cm^{-3} were found. The lifetime of the minority carriers was found to be $0.1-1$ μs, depending on the pressure during evaporation. It is suggested, that the interfaces between layers and substrates possess the worst structural properties and that the electrical properties of $p-n$ junctions are much more characteristic of the layer quality than the results of the proper structural investigations.

Es wurden Ge-Schichten auf p-leitende Substrate aufgedampft bei verschiedenen Drucken zwischen 3×10^{-5} und 1×10^{-8} torr. Der Pumpenstand war mit einer Öldiffusionspumpe versehen. Es ergaben sich folgende Aufdampfparameter: mit CP-4 geätzter (111)-Ge-Träger, Temperatur des Trägers 810 °C, Kondensationsrate $3-5$ μmin^{-1}, Schichtdicke $15-25$ μ. Die Struktureigenschaften wurden mit Hilfe von mikroskopischen Untersuchungen, Elektronendiffraktionsaufnahmen und Untersuchung der Ätzrate geprüft; $p-n$ Übergänge wurden durch Diffusion hergestellt. Es wurden I–V und C–V-Charakteristiken der Übergänge aufgenommen und die Speicherungszeit gemessen. Es ergaben sich Akzeptorkonzentrationen von 1×10^{18} bis 4×10^{15} cm^{-3}. Die Lebensdauer betrug $0{,}1-1$ μs und wies eine Druckabhängigkeit auf. Es wurde festgestellt, daß die Grenzschicht zwischen den Trägern und den Aufdampfschichten die unvollkommenste Struktur besitzt; weiterhin, daß die elektrischen Parameter des $p-n$ Überganges viel mehr über die Struktur dieser Grundschicht auszusagen vermögen als die üblichen Strukturuntersuchungsmethoden.

1. Introduction

Vacuum condensed Si and Ge layers are generally not of "device quality". The Ge-layers evaporated from an intrinsic source material are always of p-type containing a great number of lattice defects. The possible causes are the following:

a) In practice the evaporation takes place on a real, oxide covered substrate surface.

b) The condensation parameters lie far from those of the thermodynamic equilibrium.

c) During the condensation there is a possibility for the incorporation of the contaminations from the ambient.

Only the exact knowledge of the relationship between formation conditions and layer parameters gives the explanation of these problems.

Previous investigations clarified the effect of the layer thickness, the substrate material, the substrate temperature and the condensation rate on the layer parameters (KUROV et al.; DAVEY et al 1963, 1966; SAKAI, TAKAHASHI; SLOOPE, TILLER 1963, 1965; COURVOISIER et al., 1963; CATLIN, HUMPHRIS; HUMPHRIS, CATLIN; MIETZ; WOODMAN). The p-type character of undoped Ge-layers is not quite clear yet. The acceptor concentration of undoped layers grown by COURVOISIER, HAIDINGER, JOCHEMS and TUMMERS differ considerably from the values obtained by DAVEY and others. The effect of the system pressure is not sufficiently clarified either. There is only a slight difference, if any, in the results of high-vacuum and ultrahigh-vacuum evaporations.

Investigations made on layers evaparated at various system pressures are discussed in this paper. All the formation conditions besides the pressure were kept constant possibly at an optimal value. Structural investigations were performed, the characteristics of the diffused p–n-junctions developed on the layers were determined and the acceptor concentration of the layers was calculated.

2. Layer preparation

The evaporation of Ge took place in all metal vacuum systems evacuated by oil diffusion pumps. In an Edwards 12 EA system the pressure before and during the evaporation amounted to about $2-3 \times 10^{-5}$ torr. Another series of high-vacuum experiments were carried out in an ultrahigh-vacuum equipment (Balzers BA 650). Without baking the system pressure is $2-3 \times 10^{-7}$ torr, during the evaporation it amounts to $6-7 \times 10^{-7}$ torr. After outgassing the pressure is $1-2 \times 10^{-9}$ torr and during the evaporation it amounts to 1×10^{-8} torr.

The substrate was p-type Ge with a specific resistivity of 1 ohmcm. Before evaporation the samples were cut parallel to the (111) faces, lapped, polished mechanically and etched in CP-4. The substrate temperature was 810 °C.

The evaporation source was a small piece of intrinsic polycrystalline Ge melted previously on a tungsten filament and heated directly (HARASZTHY, PFEIFER). The condensation rate was $3-5\ \mu\mathrm{min}^{-1}$; the average thickness of the layers $15-25\ \mu$. The obtained layers were not homogeneous, the thickness difference between the center and the edge amonuted to about $20-30\%$.

3. Examination of the layers

The crystalline structure of the layers was investigated by electronmicroscopy, electron- and X-ray diffractometry. The defect density of the layers was determined. The microstructure was studied by structure sensitive physicochemical processes such as alloying, diffusion and etching.

In order to examine the electric properties p–n junctions were formed in the layers and the electrical characteristics of the diodes were determined.

3.1 Crystalline structure

It has been proved by both electrondiffractograms and Laue-patterns, that all the layers deposited in the vacuum range of 10^{-5}, 10^{-7} and 10^{-8} torr are well oriented monocrystals as expected. The Kikuchi-lines are the sharpest in the case of layers deposited under ultrahigh-vacuum conditions. Apart from this the layers deposited in the three mentioned pressure ranges show no difference as regards the surface and the inner part developed by step-by-step etching.

The etching in Billig etchant developed stacking faults, microtwins and dislocations. The dislocation density of the layers is rather high, $2-9\times10^5$ cm^{-2}. The dislocation distribution is not homogeneous. It is remarkable that the dislocations are accumulating at the edges and at certain preferred sites. The dislocation density in the layers deposited at 10^{-5} torr is about 4–5 times as high as in layers deposited at 10^{-8} torr. Figure 1 displays the defects of layers grown in the three pressure ranges. The defect density including the density of the stacking faults and that of microtwins amounts to 10^4-10^5 cm^{-2}.

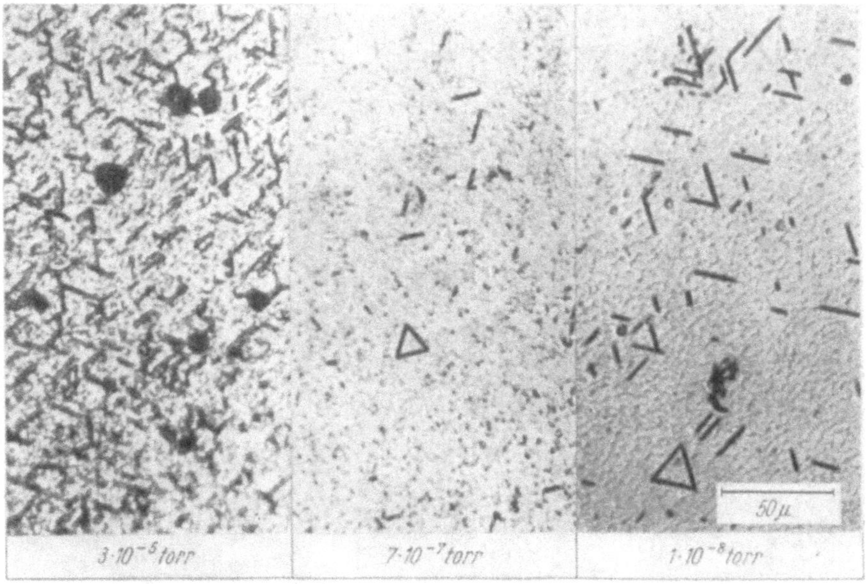

Fig. 1. Stacking faults and microtwins developed with Billig etchant in layers deposited at 3×10^{-5}, 7×10^{-7}, and 1×10^{-8} torr

This value agrees with the findings of Woodman, especially considering that the thickness of the layers is about 10 time as thick as the layers grown by Woodman, and as it can be seen on the figure the size of the stacking faults is different, proving that not each originates from the interface. The defect density (4×10^4, 6×10^4 and 2×10^5 cm^{-2}) increases with the system pressure. The change in surface morphology depending on the system pressure is even more remarkable than the difference between the defect densities.

3.2 Microstructure

Interesting results were achieved by step-by-step etching of the layers in liquid-phase etching experiments. Figure 2 shows the etching rate in the layer plotted against the distance measured from the original surface of the layer. The applied etchant, superoxole etches the Ge slice used for comparison with a uniform rate. Etching of the layers gives another result. In the bulk of the layers the etching rate is constant while on the surface the etching rate is lower. In the vicinity of the interface the etching rate is much higher than inside. This high etching rate in the vicinity of the interface agrees with the findings of Haidinger, Courvoisier and Arizumi. This behaviour of the etching rate

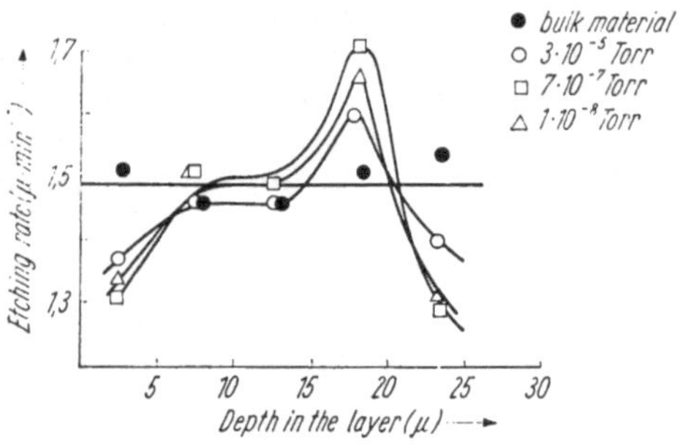

Fig. 2. Etching rate of deposited layers and bulk material in superoxole vs the distance measured from the surface of the sample

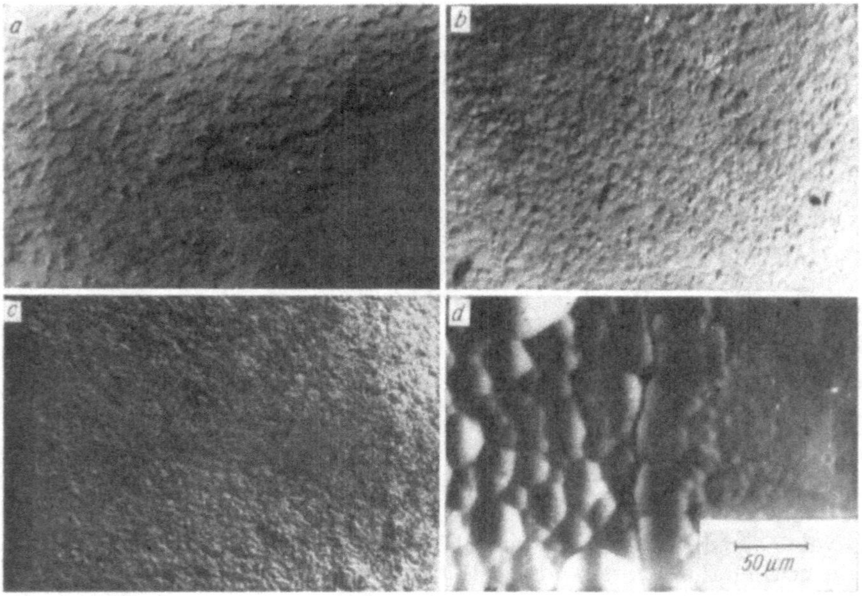

Fig. 3. a – surface of a deposited layer, b –, c – the "layer bulk", d – substrate bulk developed in superoxole

does not depend on the pressure which exists during the evaporation. Figure 3 shows the surface of the condensed layer, the structure of the "layer bulk" developed in superoxole the process as the layer becomes rough at the interface and finally a pattern is shown lying deeper than the interface. Both parts of Figure 3d show micrographs produced inside of the bulk. The structure of the bulk under the interface is characteristic for the mechanism of the development and/or for the dissolution of the interface. The other part of the micrograph is a typical pattern of Ge etched in superoxole. Vacuum-deposited layers are of an unusual character taking into account not only etching but alloy forming processes as well. Surfaces obtained by vacuum condensation can hardly be wetted by Pb–Sb–Sn-type alloys. But the diffusion process in the layers is analogous to that in Ge substrates. The hole concentration of the layers is estimated by the depth of diffusion to 2×10^{15} cm^{-3}. This date is in agreement with the electrical measurements.

3.3 Diode characteristics

p–n junctions were formed by deep diffusion of Sb in layers condensed 3×10^{-5} and 1×10^{-8} torr. At the junction the concentration distribution of Sb we can consider to be linear. In order to have a qualitative knowledge about the distribution of the layer properties on the surface and also on the edges of the layers, the characteristics were determined after contacting and mesa-etching, before they were sliced.

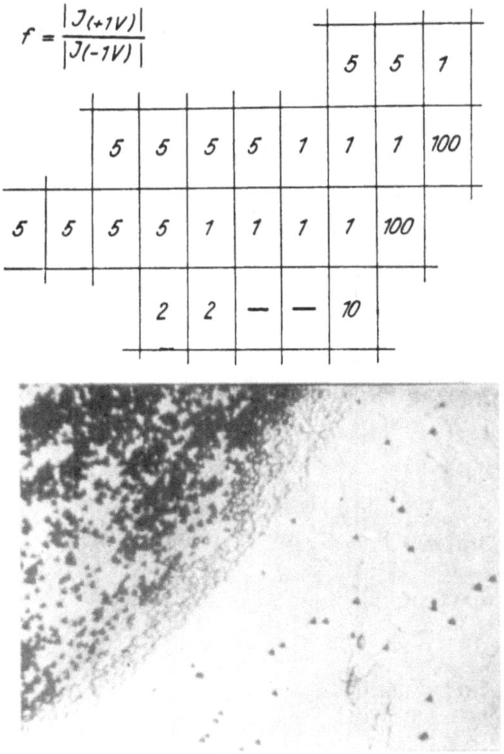

Fig. 4. Upper part: arbitrary values characteristic for the rectifying properties of diodes made on a part of the layer. Lower part: dislocation distribution of the same layer

Quantitative evaluation was made by the usual method on the basis of experimental results on capsuled diodes. In Figure 4 arbitrary values are plotted which are characteristic for the rectifying properties of the diodes on a given part of the layer. In the lower part of the figure the dislocation distributions of the layer, of the edge and of the substrate, are shown. The diodes formed inside the layer are rather similar, their rectifying effect is much worse than of those formed on the substrate. At the edge of the layer the diodes do not rectify, they have ohmic characteristics. This phenomenon, and the fact, that the edge of the layer is so inhomogeneous that even dislocations can not be found in it, show the importance of the interface again. The edge of the layer is obviously analogous to the interface.

The room temperature characteristics of the capsuled diodes are similar and agree with the characteristics formed on the substrate. This gives a possibility to determine the carrier concentration in the epitaxial layer and the lifetime of the minority carriers.

3.3.1 Capacitance-voltage characteristics

Hole-concentration was determined by capacitance-voltage characteristics. The hole-concentration at the p–n junction was found to be $1-4\times10^{15}$ cm^{-3}. This agrees considerably better with the results of COURVOISIER, HAIDINGER, JOCHEMS and TUMMERS than with the results of DAVEY. The average hole-concentration of a layer condensed at 3×10^{-5} torr is twice as much as in layers condensed at 1×10^{-8} torr.

Further conclusions may also be drawn by the investigation of the capacitance-voltage characteristics of the diodes. Figure 5 shows l/c^{-3} plotted against V. The curve intercepts the V axis at the point V_i. The intercepted voltage (V_i)

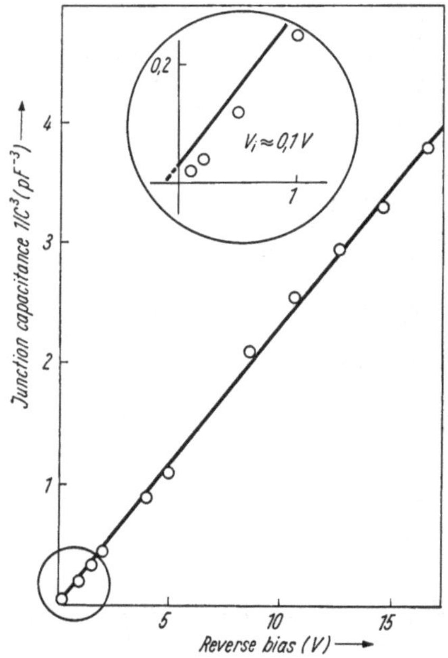

Fig. 5. l/c^{-3} vs reverse bias

is a few tenth of volts smaller than the value of the built-in voltage (V_d). According to the CHANG model the difference can be explained by the partial ionization of the impurities.

Consequently, in vacuum condensed Ge layers the impurity levels lie deeper than usually.

3.3.2 Storage-time measurement

The effective lifetime of the minority carriers can be determined by the storage-time measurement, this means, by the transient measurement of the voltage remaining after the interruption of the forward current on the diodes. The effective lifetime shows a significant difference in the layers condensed at 3×10^{-5} and 1×10^{-8} torr. The effective lifetime at the higher pressure is about 0,1 μs while at the lower pressure it is about 0,4—1 μs.

The question arises whether the effective lifetime is determined by the recombination in the "layer bulk", or if it is influenced by the recombination processes at the interface and, whether the latter is dominant.

The problem should be investigated separately for samples condensed at 10^{-5} and 10^{-8} torr because of the great difference between the effective lifetime values. On the basis of effective lifetime values of 10^{-8} torr samples and the model of the narrow base diodes, it could be proved that the effective lifetime is determined by the recombination at the interface even if the mobility in the layer is one tenth of the mobility in the bulk. According to DAVEY no more decrease of the mobility value is to be taken into account.

The effective lifetime in layers condensed at 3×10^{-5} torr is so low that the interface does not influence it; the bulk-recombination is dominant.

3.3.3 Current-voltage characteristics

At low current density the forward current characteristics are in agreement with the exponential law. There is a significant difference between the breakdown voltage and the reverse current characteristics measured at 10^{-5} torr and 10^{-8} torr. Figure 6 shows typical reverse current characteristics of the diodes formed in layers condensed in the above mentioned vacuum ranges. The average saturation current at 10^{-8} torr is 2×10^{-3} A cm^{-2}. 10^{-5} torr diodes have not good saturation properties because of surface leakage current. The saturation current was found to be about $5-10 \times 10^{-2}$ A cm^{-2}. The typical breakdown voltages measured at high current densities were in the higher and lower vacuum 40 and 25 V respectively.

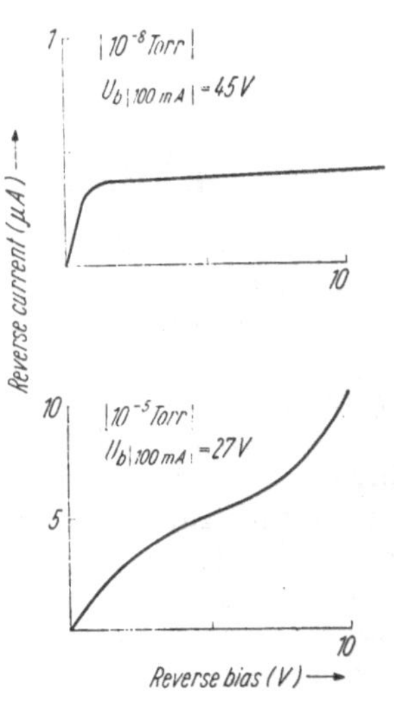

Fig. 6. Typical saturation characteristics of diodes formed on layers deposited at 1×10^{-8} and 3×10^{-5} torr

Here we would like to point out that we could calculate the theoretical values of breakdown voltage from the concentration gradient according to the model of SZE and GIBBONS. The theoretical breakdown voltages amount to 90 and 75 V respectively. The measured lower breakdown voltages refer to crystal defects in the condensed layer. The difference between the theoretical and mea sured breakdown voltage is higher in layers grown at 10^{-5} torr than in those obtained at 10^{-8} torr. This difference proves that the crystal structure of a layer grown in a lower vacuum is more imperfect.

Table 1

Parameters characterising deposited Ge layers

System pressure (Torr)		$3 \cdot 10^{-5}$	$7 \cdot 10^{-7}$	$\leqq 1 \cdot 10^{-8}$
$R = \frac{\text{Ge atoms deposited}}{\text{impinging gas molecules}}$		23	1000	$\geqq 70000$
Crystalline structure		highly oriented kikuchi lines		
Microstructure	dislocation density (cm^{-2})	$9 \cdot 10^{5}$	$3 \cdot 10^{5}$	$2 \cdot 10^{6}$
	defect density (stackingfaults + microtwins) (cm^{-2})	$2-7 \cdot 10^{5}$	$5-6 \cdot 10^{4}$	$4-6 \cdot 10^{4}$
Electrical properties	carrier type	P	P	P
	carrier concen- tration (cm^{-3})	$2-4 \cdot 10^{15}$		$1-3 \cdot 10^{15}$
	carrier lifetime (diffused diodes) (μsec)	0,07 – 0,09		0,4 – 1
	breakdown voltage (diffused diodes) (V)	20 – 30		35 – 54
	saturation current (diffused diodes) ($A \cdot cm^{-2}$)	$5-10 \cdot 10^{-2}$		$1-3 \cdot 10^{-3}$

4. Conclusions

Table 1 summarizes the results of experiments and measurements. Evaporations performed at different pressures, produce layers of different properties. Comparing a homogeneous Ge sample with a structure containing a layer deposited on an oxide covered surface the most remarkable difference lies not in the layer itself but in the presence of the interface. This was proved by the etching processes, by the electrical characteristics of the diodes formed on thin layers. The properties of the "layer bulk" can be compared with the properties of the substrate as it was shown by structural, diffusion, etching investigations and by characteristics of diodes.

From the mentioned investigations, the diode measurements proves to be most characteristic for the difference between substrate and layer bulk, and between the layer bulks, produced at different pressures.

Acknowledgements are due to Dr. G. SZIGETI, Director of the Institute and to Dr. G. GERGELY for their interesting discussions and kind assistance, further to Mrs. M. ANDRÁSI for having prepared the diffused diodes with great care.

Literature

ARIZUMI, T., NISHINAGA, T., KAKEHI, M.: Jap. J. appl. Physics **7**, 468 (1968)
CATLIN, A., HUMPHRIS, R. R.: Int. Symp. Clausthal, Göttingen 1965
CHANG, Y. F.: Solid-State Electronics **10**, 281 (1967)
COURVOISIER, J. C., HAIDINGER, W., JOCHEMS, P. J. W., TUMMERS, L. Y.: Solid-State Electronics **6**, 265 (1963)
DAVEY, J. E.: Appl. Physics Letters **8**, 164 (1966)
DAVEY, J. E., TIERNAN, R. J., PANKEY, T., MONTGOMERY, M. D.: Solid-State Electronics **6**, 205 (1963)
HAIDINGER, W., COURVOISIER, J. C.: Le Vide **104**, 141 (1963)
HARASZTHY, E., PFEIFER, J.: to be published
HUMPHRIS, R. R., CATLIN, A.: Solid-State Electronics **8**, 957 (1965)
KUROV, G. A., SEMILETOV, S. A., PINSKER, S. G.: Kristallografiya **2**, 59 (1957)
MIETZ, J.: Kristall u. Technik **3**, 177 (1968)
SAKAI, Y., TAKAHASHI, K.: Jap. J. appl. Physics **2**, 629 (1968)
SLOOPE, B. W., TILLER, C. O.: J. appl. Physics **36**, 3174 (1965)
SLOOPE, B. W., TILLER, C. O.: Trans. 10th Nat. Vac. Symp., New York, **339** (**1963**)
SZE, S. M., GIBBONS, G.: Appl. Physics Letters **8**, 111 (1966)
WOODMAN, T. P.: Thin Solid Films **2**, 173 (1968)

(Received April 28, **1969**)

Authors' address:

J. PFEIFER
L. GUTAI
E. Sz. HARASZTHY
Research Institute for Technical
Physics of the Hungarian Academy of Sciences
Budapest IV, Ujpest, Fóti ut 56, Hungary

Kristall und Technik | 5 | 1 | 1970 | 95–99

E. Schippel

VEB Werk für Bauelemente der Nachrichtentechnik
„Carl von Ossietzky“, Teltow

Elektronenmikroskopische Untersuchung zur Struktur aufgedampfter Nickel/Chrom-Schichten[1])

Zur elektronenmikroskopischen Untersuchung von auf Glas und keramischen Substraten aufgedampften Ni/Cr-Schichten wurde Zellin als wasserlösliche Abdruckmatrize verwendet.

Zur Erkennung der Schichtstruktur war eine Anätzung der Oberfläche erforderlich, die aufgrund der Passivierung des vorliegenden Schichtmaterials besonders schwierig ist. Die getemperte Legierungsschicht zeigt im Elektronenmikroskop zwei verschiedene Phasen. Die Ni-reiche, kubisch kristalline Phase mit ausgeprägten idiomorphen Kristallen ist eingelagert in einer „schotterartigen“ Cr-reichen Phase.

For an electron-microscopical examination of Ni/Cr-layers evaporated on glass and ceramic substrates, Zellin was used as water-soluble impression matrix.

To realize the layer structure, an etching of the layer surface was necessary which was rather difficult in view of the passivating of this layer material. The annealed alloy film shows two different phases in the electron microscop. The Ni-rich, cubical crystalline phase with markedly idiomorphic crystals is embedded in a "ballast-like" Cr-rich phase.

Elektronenoptische Untersuchungen sind an aufgedampften Ni/Cr-Schichten von Campbell, Hendry; Bicknell, Blackburn, Campbell, Stirland; Blackburn, Blicknell durchgeführt worden. Die Untersuchungen erfolgten mittels Elektronenbeugung und Durchstrahlung von auf NaCl-Substraten aufgedampften Schichten. Nach diesen Untersuchungen besteht die Schicht aus Inseln einer kubisch kristallinen Phase, die in einer amorphen „Haut“ eingelagert sind (Blackburn). Die kubisch kristalline Phase ist Ni-reich und kann bis zu 30% Chrom enthalten (Bicknell, Blackburn, Campbell, Stirland), während die amorphe Phase Cr-reich ist und einen hohen Oxidanteil von Cr_2O_3 enthält. Die freitragend präparierten Schichten wurden auch getempert. Die stark diffusen Beugungsreflexe der ungetemperten Schicht gehen nach der Temperung in ein scharfes Beugungsbild über.

Aufdampfschichten aus Nickel/Chrom auf Glas und keramischen Substraten werden für Metallschichtwiderstände in der Elektrotechnik verwendet. Bei der Untersuchung der Schichteigenschaften (Schippel 1968 und 1969) wurde u. a. festgestellt, daß bei der Temperung im Vakuum ab 300 °C eine plötzliche, stark negative Wertänderung des elektrischen Widerstandes auftritt (Fig. 1). Die negative Wertänderung hält bis ca. 650 °C an. Bei höheren Temperaturen bzw. längeren Temperzeiten bei 700 °C ist wieder ein Anstieg des Widerstandswertes zu verzeichnen. Dabei steigt der Wert mit abnehmender Schichtdicke zunehmend schneller gegen unendlich.

[1]) Vorgetragen auf der Arbeitstagung für Elektronenmikroskopie in Dresden, Februar 1969.

Ziel der Untersuchungen war es, von diesen, auf den Substraten getemperten Ni/Cr-Schichten, elektronenmikroskopische Aufnahmen zur Phasenausbildung und -verteilung zu erhalten.

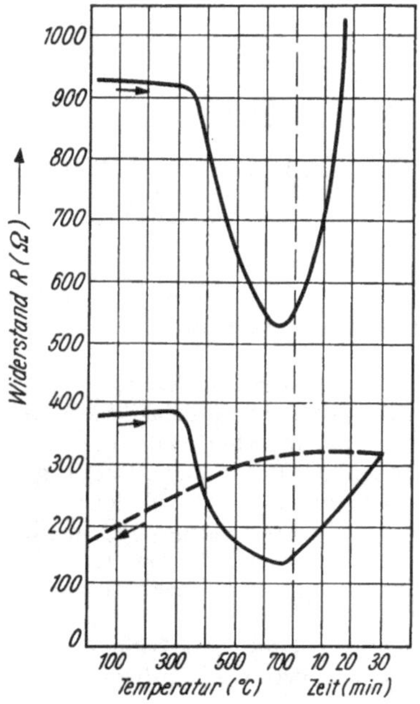

Fig. 1. Widerstandsänderungen während der Temperung von Ni/Cr-Schichten bis auf 700 °C und während der Halteperiode von 30 min auf 700 °C

Zur Herstellung der elektronenmikroskopischen Präparate wurde Zellin vom VEB Filmfabrik Wolfen verwendet. Es wurde eine gesättigte Lösung von Zellin in Wasser hergestellt und bei 80 °C homogenisiert. Die auf der Probe in einem evakuierten Exikator eingetrocknete Zellinfolie kann abgehoben und nach der Beschichtung mit Kohle in Wasser (auf der Oberfläche schwimmend) wieder gelöst werden. Die Folienstücke von ca. 4 ×4 mm² Größe wurden zur Auflösung zweimal für 20 min auf kaltes Wasser und abschließend für 5 bis 10 min auf heißes Wasser gebracht. Danach hatte sich die Zellinmatrize vollständig gelöst und der Kohlefilm konnte zur Schrägbeschattung mit Chrom auf die Objektnetzträger übertragen werden.

Elektronenmikroskopische Aufnahmen von der getemperten Ni/Cr-Schicht zeigen, daß die Schichtoberfläche topographisch unverändert bleibt. Zur Erkennung der Schichtstruktur nach morphologischen Merkmalen ist eine Anätzung der Schicht erforderlich. Die Ätzung ist aufgrund der Passivierung dieses Schichtmaterials eines der schwierigsten Probleme der elektronenmikroskopischen Untersuchung. Am geeignetsten erwiesen sich HCl bzw. HCl-haltige Ätzmittel.

Figur 2 zeigt die Aufnahme einer ungetemperten Ni/Cr-Schicht. Die Passivierungshaut ist noch vorhanden. Die Ätzung erfolgte hier offenbar unterhalb der Passivierungshaut, die an einigen tieferen Ätzstellen nachträglich bei der Probentrocknung aufgerissen ist. Die Resistenz der Passivierungshaut gegen HCl nimmt mit steigendem Ni-Gehalt der Schicht ab.

Aufnahmen von der angeätzten getemperten Ni/Cr-Schicht lassen deutlich zwei Phasen in der Legierungsschicht erkennen (Fig. 3 bis 6). Die kubisch

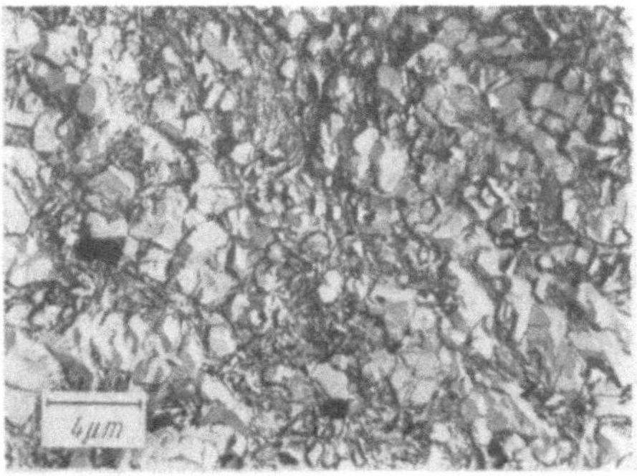

Fig. 2. Ungetemperte angeätzte Ni/Cr-Schicht mit aufliegender Passivierungshaut (Ätzmittel: HCl)

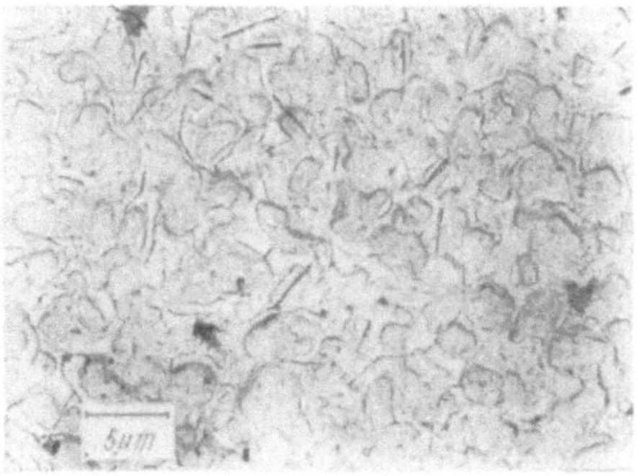

Fig. 3. Phasenverteilung in einer auf Glas aufgedampften und getemperten Ni/Cr-Schicht (getempert 30 min bei 450 °C, Ätzmittel: HCl)

Fig. 4.

Fig. 5.

Die kubisch kristalline Ni-reiche Phase in der ausgeprägten idiomorphen Form

Fig. 6. Häufungsstellen der Phasen in Mikrobereichen

kristalline Ni-reiche Phase mit ausgeprägten idiomorphen Kristallen ist eingelagert in einer „schotterartigen“ Cr-reichen Phase, anscheinend plättchenförmiger Ausbildung. Die kubische Phase zeigt orientierte Verwachsungen. Die Größe der Kristalle ist zum Teil sehr unterschiedlich. Die Verteilung der beiden Phasen ist in Mikrobereichen nicht gleichmäßig (Fig. 6). Es gibt Häufungsbereiche der einen oder der anderen Phase, dabei scheinen Fremdpartikel (Staubteilchen) nicht ohne Einfluß zu sein. Innerhalb größerer Schichtbereiche liegt jedoch im allgemeinen eine statistische Verteilung der beiden Phasen vor (Fig. 3).

Aufnahmen von anderen Stellen der gleichen Probe, die vom Ätzmittel weniger stark angegriffen worden sind, zeigen Figur 7 und 8. In der ursprünglich glatten Schichtoberfläche sind hier gerade die Kristallkonturen erkennbar. Diese Aufnahmen der getemperten Schicht zeigen, daß die Umwandlungsprozesse und das

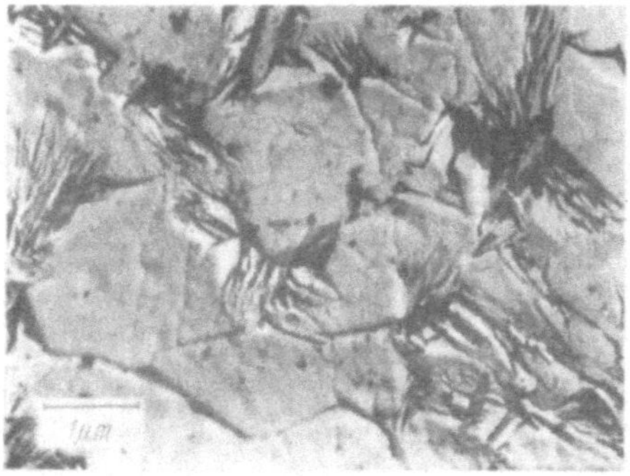

Fig. 7

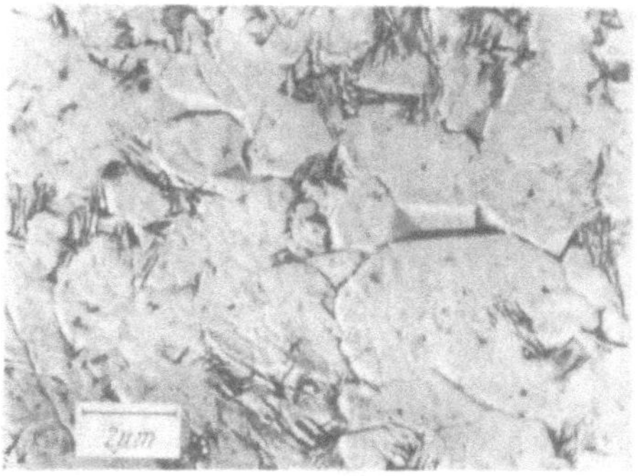

Fig. 8.

Durch schwächere Ätzung erzeugte Kristallkonturen in der ursprünglich glatten Schichtoberfläche der getemperten Ni/Cr-Schicht

Kristallwachstum im Schichtinnern die Schichtoberfläche nicht verändern. Dies fanden auch Blicknell, Blackburn, Campbell und Stirland bei ihren Untersuchungen: "It is most noticeable that since all areas retain the topography of the film as originally deposited, the phase changes have probably occurred with no mass-transport." Diese Erscheinung der Ni/Cr-Schicht muß im Zusammenhang mit der Passivierung gesehen werden. Eine Kristallausbildung entsprechend den Figuren 3 bis 6 ist ohne Massentransport nicht vorstellbar. Für das Kristallwachstum stellt die Passivierungshaut auf der Schichtoberseite eine ebensolche Wachstumsschranke dar, wie das Substrat auf der Schichtunterseite. Kristallwachstum und Strukturänderungen während der Temperung können somit nur innerhalb der Schicht erfolgen, ohne daß diese Prozesse die Passivierungshaut verändern oder durchstoßen.

Die Schichtaufdampfung für diese Untersuchungen erfolgte durch Sublimation eines widerstandserhitzten Ni/Cr-Drahtes (2 mm ∅) von der Zusammensetzung Ni80/Cr20. Um die Schichtzusammensetzung trotz des unterschiedlichen Abdampfens von Nickel und Chrom konstant zu halten, wurde kurzzeitig (5 min) intensiv von einer Drahtwendel abgedampft. Die Schichtzusammensetzung beträgt nach röntgenspektroskopischer Bestimmung ca. 40Ni/60Cr (Schippel 1968).

Die starken strukturellen Änderungen in der Ni/Cr-Schicht finden auch ihren Ausdruck im Temperaturkoeffizienten (Tk) des elektrischen Widerstandes, der gegenüber der ungetemperten Schicht mit $\pm < 100 \times (10^{-6}/°C)$ auf positive Werte von 500 bis $600 \times (10^{-6}/°C)$ ansteigt. Übereinstimmend mit den Beugungsuntersuchungen von Bicknell, Blackburn, Campbell und Stirland; Blackburn; Bicknell ist festzustellen, daß die geringen Widerstandsänderungen bei der Temperung bis 300 °C dem positiven oder negativen Tk entsprechen. Die plötzliche, stark negative Wertänderung des elektrischen Widerstandse ab 300 °C wird durch das spontan einsetzende Kristallwachstum der Ni-reichen Phase hervorgerufen. Das weitere Absinken des Wertes bis 650 °C ist durch die zunehmende Kristallisation dieser Phase bedingt. Die nachfolgende stark positive Wertänderung oberhalb 700 °C wird durch die nun einsetzende Kristallisation der Cr-reichen Phase verursacht, die bis dahin unverändert bleibt. Es wurde eine Abhängigkeit der Widerstandsänderungen von der Schichtzusammensetzung bei der Temperung in der Art gefunden, daß mit steigendem Anteil der Cr-reichen Phase in der Schicht die negativen Änderungen bis 650 °C geringer und der positive Wertanstieg oberhalb 650 °C zunehmend steiler wird.

Literatur

BICKNELL, R. W.: J. appl. Physics **17**, 775–781 (1966)
BICKNELL, R. W., BLACKBURN, H., CAMPBELL, D. S., STIRLAND, D. J.: Microelectr. and Reliability **3**, 61–64 (1964)
BLACKBURN, H.: Vakuum-Techn. **14**, 200 (1965) (Tagungsbericht)
CAMPBELL, D. S., HENDRY, B.: J. appl. Physics **16**, 1719–1725 (1965)
SCHIPPEL, E.: Nachrichtentechn. **18**, 434–437 (1968)
SCHIPPEL, E.: Exp. Techn. Physik **17**, 175–182 (1969)

(Eingegangen am 25. März 1969)

Anschrift des Verfassers:

Dr. Erhard SCHIPPEL
Werk für Bauelemente der Nachrichtentechnik
„Carl von Ossietzky"
153 Teltow
Ernst-Thälmann-Str. 10

Kristall und Technik | 5 | 1 | 1970 | 101–107

L. O. Meleško

Minsk, UdSSR

Die Abhängigkeit der Aktivierungsenergie des Kristallwachstums von der Temperatur

Es wird eine Methode zur Bestimmung der Temperaturabhängigkeit der Aktivierungsenergie der Grenzschicht-Selbstdiffusion im Bereich des abfallenden Teils des Verlaufs der Funktion $v = f(\Delta T)$ (v ist die lineare Kristallwachstumsgeschwindigkeit, ΔT die Unterkühlung der Schmelze) beschrieben, ein entsprechender analytischer Ausdruck abgeleitet und experimentelle Daten für Antipyrin, Benzophenon und Kodein mitgeteilt.

Предлагается методика определения температурной зависимости энергии активации граничной самодиффузии по ниспадающей ветви кривой $v = f(\Delta T)$ (v-линейная скорость роста кристаллов, ΔT-переохлаждение расплава).
Получены аналитические выражения температурной зависимости энергии активации граничной самодиффузии молекул антипирина, бензофенона и кодеина.

1. Einführung

Beim Studium des Mechanismus des Kristallwachstums hat die Aktivierungsenergie der Grenzschicht-Selbstdiffusion, d. h. die Energie, die für den Übergang der Teilchen aus einer zeitweiligen Gleichgewichtslage in der Flüssigkeit in die Gleichgewichtslage an der Kristalloberfläche erforderlich ist, eine entscheidende Bedeutung. Dieser Übergang wird bedingt durch den „kollektiven" Umbau der vorhandenen Teilchenanordnung in eine Anordnung, bei der sich günstige Bedingungen für den Übergang eines Moleküls (Atoms) in den neuen Zustand ergeben. Als Quelle der zusätzlichen Energie, die einen solchen Übergang sichert, erscheint der Wärmeaustausch der Teilchen. Die Energie, die das Molekül besitzen muß, damit dieser Elementarakt vor sich geht, wird eben gerade durch die Energie der lokalen Störung der Teilchenanordnung bestimmt. Bei den Berechnungen wird die begründete Annahme gemacht, daß am Elementarakt nur das wandernde Teilchen teilnimmt, während die Umgebung gleichzeitig unverändert bleibt. In diesem Falle, d. h. bei einer sozusagen feststehenden Umgebung, erhält die Aktivierungsenergie vereinfacht den Charakter der Höhe der Potentialschwelle, die den Übergang eines Moleküls von einer Position in eine andere verhindert.

Die derzeitigen Theorien bestimmen nicht die Größe der Aktivierungsenergie, sondern machen nur allgemeine Angaben, daß sie bei Kristallen von der Gitterenergie bzw. bei Flüssigkeiten von der Verdampfungswärme abhängt. Da diese Größen Funktionen der Temperatur sind, ist eine Temperaturabhängigkeit der Aktivierungsenergie möglich. Die Erfahrung zeigt, daß man die Aktivierungsenergie für kristalline Körper in grober Annäherung in einem kleinen Temperaturintervall als konstant ansehen kann, während bei Flüssigkeiten eine solche Annahme offensichtlich nicht vertretbar ist (PANČENKOV; KOBEKO, KUVŠINSKIJ, ŠIŠKIN; ŠIŠKIN).

Bei der Analyse des Mechanismus des Kristallwachstums aus der Schmelze nimmt man an, daß die Aktivierungsenergie u gewöhnlich gleich der Aktivierungsenergie der Selbstdiffusion im Volumen der Schmelze ist, wobei vorausgesetzt wird, daß der Übergang des Atoms aus der zeitweiligen Gleichgewichtslage in der Flüssigkeit in die Gleichgewichtslage an der Kristalloberfläche einen Diffusionsübergang auf eine Zwischenatomlage darstellt (JANTSCH; HILLIG, TURNBULL).

Wie in einer Reihe von Arbeiten (GIBBS; CAHN, HILLIARD) gezeigt wird, ändert sich an der Grenze Kristall-Schmelze die Dichte eines Stoffes erheblich (in einer Distanz von nur einem Atomdurchmesser), aber stetig in den Grenzen einiger zwischenatomarer Abstände. In diesem Gebiet ändern sich die thermodynamischen Parameter ständig von Werten, die für die eine Phase charakteristisch sind, zu Werten einer arteigenen anderen Phase. Solche Phasengrenzen nennt man diffus. Die Eigenschaften der Grenzschicht sind eng mit den Eigenschaften der Kristalle und der Flüssigkeit verbunden. Die Verwaschung der Phasengrenze ist bedingt durch das nahe der Kristalloberfläche vorhandene Kraftfeld, dessen Struktur vom Bau der Moleküle, vom mikrogeometrischen Bau der äußeren und inneren Oberfläche der festen Phase, aber auch von Defekten aller Art (Rissen, Mosaikstrukturen, Hohlräumen, Fremdstoffeinschlüssen, Versetzungen u. a.) bestimmt wird. Hierbei ist natürlich anzunehmen, daß die Struktur des Kraftfeldes, das sich um die Festkörperoberfläche ausbildet, nicht nur von den Oberflächenschichten, sondern auch von den inneren, die sich in einiger Tiefe innerhalb des Kristalls befinden, bestimmt wird.

Es muß angenommen werden, daß Gradient und Aussehen des Potentialfeldes an der Kristalloberfläche und folglich auch die Breite des Diffusionsfeldes, von der Intensität der Wärmebewegung abhängen. Je niedriger die Temperatur ist, desto stärker tritt der periodische Charakter der Kristalloberfläche hervor und desto größer sind der Gradient des Kraftfeldes und die Breite des Diffusionsfeldes. Der Wachstumsmechanismus der Kristalle wird wesentlich bestimmt durch die Größe dieses Gradienten und durch die Struktur dieses Kraftfeldes, die kondensierte Phase sowie durch die Art ihres Zusammenwirkens mit den Atomfeldern und den Molekülen der unterkühlten Schmelze.

Wenn sich ein Molekül der Schmelze einer Kristallfläche nähert, so gelangt es in den Wirkungsbereich des durch die Oberflächenatome und -moleküle erzeugten, gerichteten Kraftfeldes. Dabei geht in den polaren Molekülen eine Umverteilung der Ladungen vor sich. In dipolaren Gruppen beginnen Drehmomente zu wirken. Unter diesen Bedingungen verliert das Molekül in erheblichem Grade seine Beweglichkeit im Raum, und das um so mehr, je geringer die Temperatur der Schmelze ist. Daher orientieren sich die Moleküle schon in einiger Entfernung von der Kristalloberfläche entsprechend der Richtung des Feldes, und unter der Einwirkung des Gradienten des Kraftfeldes wandern sie beschleunigt in Richtung der Kristallgrenze. In größerer Nähe der Kristalloberfläche setzen die Abstoßungskräfte dem Eindringen der Moleküle ins Gitter Widerstand entgegen. Indem sie einen Teil ihrer kinetischen Energie auf das Kristallgitter übertragen, wird das Kräftegleichgewicht zwischen Anziehung und Abstoßung hergestellt. Wenn sich das Molekül mit seinem nichtpolaren Ende der Kristallfläche nähert oder wenn es in den Bereich des niederen Potentials des Kraftfeldes gelangt, dann kann es unter Einwirkung der Abstoßungskräfte eine Beschleunigung in entgegengesetztem Sinne erfahren und sich wieder in die Schmelze entfernen. Moleküle, die sich auf der Kristalloberfläche befinden, bilden eine zweidimensionale Oberflächenschicht, die eine verhältnismäßig hohe Beweglichkeit besitzt. Das Vorhandensein einer Oberflächenwanderung der Moleküle des Benzophenons wurde in der Arbeit von Volmer festgestellt. Durch die Wanderung sammeln sich die Moleküle an den Stellen des größten Potentials an und bilden Keimzentren oder Wachstumsstufen in der beweglichen Oberflächenschicht. Nach deren Entstehung verlangsamt sich die Desorption, da an der Oberfläche zusätzliche Zentren mit erhöhtem Potential entstehen.

Auf diese Weise befinden sich die Moleküle der Übergangsschicht, die die physikalische Grenze zwischen den Phasen bildet, im Kraftfeld des Kristalls, und selbstverständlich ist ihre Aktivierungsenergie verschieden von der Aktivierungsenergie der Selbstdiffusion im Volumen der Schmelze oder festen Phase in beträchtlicher Entfernung von der Phasengrenze.

So wird in der Arbeit von Laudise gezeigt, daß die Aktivierungsenergie des Wachstums der Basisfläche des Quarzes gleich 20 ± 1 kcal/Mol ist, während die Aktivierungsenergie der Selbstdiffusion in der Flüssigkeit nur einigen kcal/Mol entspricht. Daher erfordert die weitere Ausarbeitung der Theorie des Kristallwachstums die Kenntnis genauer Werte der Aktivierungsenergie der Moleküle in der Diffusionsschicht.

2. Methode zur Bestimmung der Temperaturabhängigkeit der Aktivierungsenergie der Phasengrenz-Selbstdiffusion

Die Untersuchung erfolgt im Bereich der experimentell ermittelten Kurve der Funktion $v = f(\Delta T)$, und zwar im Bereich des abfallenden Teiles der Kurve.

Die experimentellen Werte der Wachstumsgeschwindigkeit der Kristalle (Meleško 1964, Fig. 1) tragen wir in ein Koordinatensystem mit den Achsen

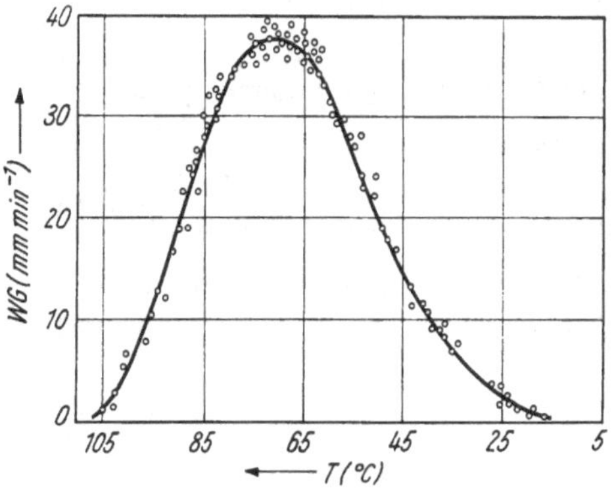

Fig. 1. Änderung der Wachstumsgeschwindigkeit der Kristalle des Antipyrins mit der Temperatur

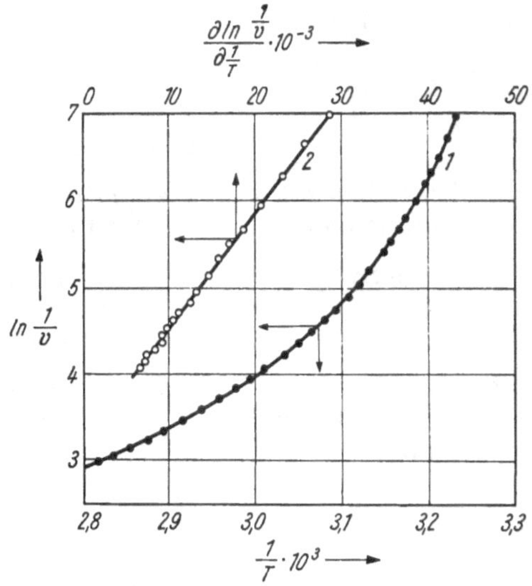

Fig. 2. Die Abhängigkeiten $\ln 1/v$ von $1/T$ beim Antipyrin (Kurve 1) und $\frac{\partial \ln 1/v}{\partial\, 1/T}$ von $\ln 1/v$ (Gerade 2)

$\ln 1/v$ und $1/T$ ein und erhalten Kurve 1 in den Figuren 2, 3 und 4. Man sieht, daß sich für Antipyrin ($C_{11}H_{12}N_2O$) (Fig. 2, Kurve 1), Benzophenon ($C_{13}H_{10}O$) (Fig. 3, Kurve 1) und Kodein ($C_{18}H_{21}NO_3$) (Fig. 4, Kurve 1) die Abhängigkeit $\ln 1/v = f(1/T)$ durch die gegen die Abszisse gekrümmten Kurven ausdrückt,

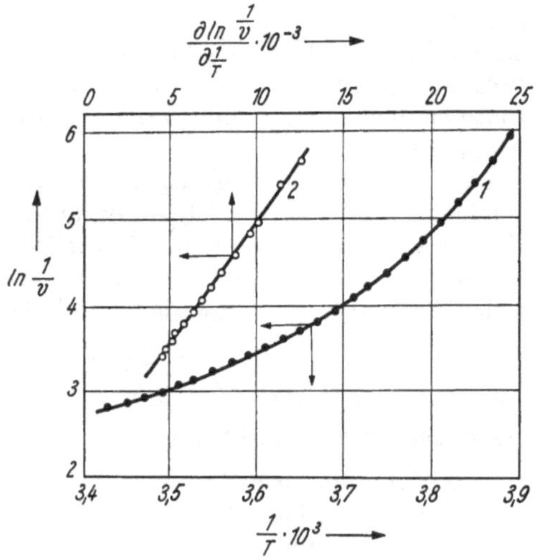

Fig. 3. Die Abhängigkeiten $\ln 1/v$ von $1/T$ beim Benzophenon (Kurve 1) und $\frac{\partial \ln 1/v}{\partial 1/T}$ von $\ln 1/v$ (Gerade 2)

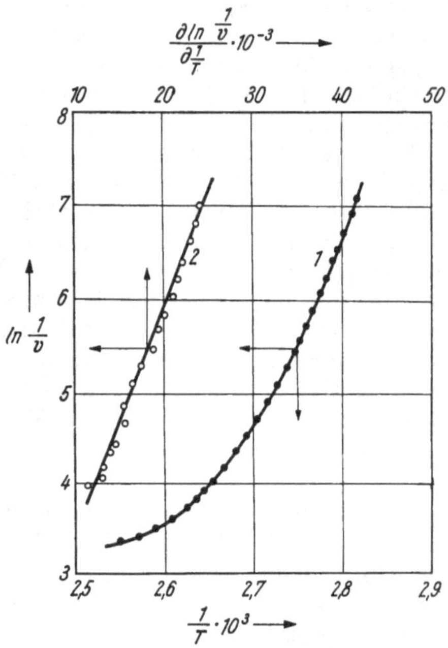

Fig. 4. Die Abhängigkeit $\ln 1/v$ von $1/T$ beim Kodein (Kurve 1) und $\frac{\partial \ln 1/v}{\partial 1/T}$ von $\ln 1/v$ (Gerade 2)

deren Neigung in jedem Punkte $\frac{\partial \ln 1/v}{\partial(1/T)}$ entspricht. Zur Bestimmung der funktionellen Abhängigkeit $u = u(T)$ finden wir die Ableitungen durch die Tangenten in bestimmten Punkten. Der Tangens der Winkel dieser Tangenten mit der Abszisse ergibt die gesuchten partiellen Differentialquotienten $\frac{\partial \ln 1/v}{\partial(1/T)}$. Die erhaltenen Daten werden in ein Koordinatensystem mit den Achsen $\frac{\partial \ln 1/v}{\partial(1/T)}$ und $\ln 1/v$ eingetragen (Fig. 2, 3, 4, Gerade 2). Man sieht, daß sich die experimentellen Punkte in den Figuren 2, 3 und 4 befriedigend in eine Gerade einfügen, was durch

$$\frac{\partial \ln 1/v}{\partial(1/T)} = a + b \cdot \ln \frac{1}{v} \tag{1}$$

entsprechend dargestellt werden kann. Durch Integration finden wir

$$\frac{1}{b} \ln\left(a + b \cdot \ln \frac{1}{v}\right) = \frac{1}{T} + \frac{1}{b} \ln c\,, \tag{2}$$

worin c die Integrationskonstante ist. Hieraus folgt

$$v = v' \cdot \exp\left(\frac{a}{b}\right) \cdot \exp\left[-\frac{c}{b} \exp\left(\frac{b}{T}\right)\right]. \tag{3}$$

Vergleicht man die empirische Formel (3) mit der theoretischen der linearen Wachstumsgeschwindigkeit der neuen Phase (AKULOV; LJUBOV), die aus den allgemeinen Beziehungen der Thermodynamik irreversibler Prozesse folgt:

$$v = A \, \Delta T \exp\left(-\frac{u(T)}{RT}\right), \tag{4}$$

so findet man

$$u(T) = RT\left[\frac{c}{b} \exp\left(\frac{b}{T}\right) + \ln \Delta T + \ln A_0\right], \tag{5}$$

worin $A_0 = \frac{A}{v' \exp(a/b)}$; $v' = 1 \frac{\text{cm}}{\text{s}}$ (Umrechnungsfaktor). Wenn man die Formel (3) mit den analogen Ausdrücken der Arbeit von ŠIŠKIN vergleicht, dann erhält man die Aktivierungsenergie der entsprechenden Mechanismen des Keim-, Spiral- und Normalwachstums.

Als eine der möglichen Methoden zur Bestimmung der Konstanten A_0 kann die folgende empfohlen werden:

Die Gleichung (4) schreiben wir in der Form

$$\ln \frac{v}{\Delta T} = \ln A - \frac{u(T)}{RT}. \tag{6}$$

Durch Differentiation von (6) nach $1/T$ findet man

$$\frac{\partial \ln \frac{v}{\Delta T}}{\partial \left(\frac{1}{T}\right)} = \frac{1}{R}\left(T \frac{\partial u}{\partial \frac{1}{T}} - u\right). \tag{7}$$

Die Versuchsdaten sind als Abhängigkeit $\ln v/\Delta T$ von $1/T$ graphisch dargestellt (Fig. 5). Im Bereich des abfallenden Teiles der Kurve, wo diese nur unbedeutend von der Geraden abweicht, ermittelten wir die Ableitung $\frac{\partial \ln v/\Delta T}{\partial \ln 1/T}$ graphisch, die uns auch den annähernden Wert u/R angibt. Wenn man u bestimmt hat, lassen sich aus Gleichung (5) A_0 und A $(A = A_0 v' \exp(a/b))$ errechnen. So erhielten wir für Antipyrin $A_0 = 0{,}25 \cdot 10^{10}$; $A = 1{,}75 \cdot 10^{8}$. Die Konstanten a und b

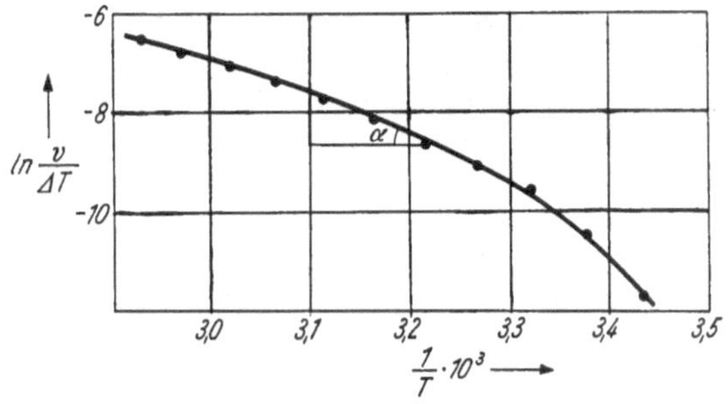

Fig. 5. Die Abhängigkeit $\ln v/\Delta T$ von $1/T$ für Antipyrin

die in (1)—(5) enthalten sind, finden wir graphisch aus den Kurven der Figuren 2, 3 und 4 (Gerade 2). Die Konstante c berechnen wir aus (2), indem wir die für a und b, und auch v gefundenen Werte bei der entsprechenden Temperatur einsetzen:

$$c = \frac{a + b \cdot \ln \frac{1}{v}}{\exp\left(\frac{b}{T}\right)} . \tag{8}$$

Für die von uns untersuchten Stoffe ergaben sich für

Antipyrin $a = -20 \cdot 10^3$; $b = 7523$; $c = 2{,}8688 \cdot 10^{-7}$

Benzophenon $a = -8 \cdot 10^3$; $b = 3636$; $c = 9{,}4 \cdot 10^{-3}$

Kodein $a = -15 \cdot 10^3$; $b = 4189$; $c = 8{,}95 \cdot 10^{-2}$.

Auf diese Weise erhöht sich entsprechend der Formel (5) die Aktivierungsenergie der Grenzschicht-Selbstdiffusion mit abnehmender Temperatur.

3. Deutung der Ergebnisse

Eine der möglichen Erklärungen für die gefundene Gesetzmäßigkeit kann die folgende sein:

Bei Erniedrigung der Temperatur der Flüssigkeit entstehen neue intramolekulare Bindungen, die im kinetischen Akt zerfallen. Die Konzentration und die Natur der intramolekularen Bindungen werden durch die Größe der Aktivierungsenergie in den einzelnen kinetischen Prozessen und durch die innere Lageordnung der Moleküle bestimmt. Die Gitterenergie kristalliner Körper hängt nur wenig von der Temperatur ab. Aus diesem Grunde ändert sich die Aktivierungsenergie unbedeutend. In Flüssigkeiten hängt die Energie der intramolekularen Bindungen wesentlich von der Temperatur ab. Daher ändert sich die Aktivierungsenergie auch wesentlich mit der Temperatur. Stoffe, die sich im flüssigen Zustand befinden, nehmen bei bestimmter Temperatur und bestimmtem Druck im Kraftfeld der kondensierten festen Phase vorher nicht vorhandene Eigenschaften an. Es steht fest, daß der Grenzschicht alle Grundformen des Aggregatzustandes eigen sind: fest, flüssig und gasförmig. Dabei wird die Änderung der Parameter des Grenzzustandes von Phasenübergängen (Derjagin, Karasev, Sorin) begleitet.

Wenn man sich auf die oben dargelegten Ausführungen stützt, ist es selbstverständlich anzunehmen, daß zur Störung der strukturellen Vollkommenheit der Moleküle der Diffusionsschicht im elementaren Akt eine Aktivierungsenergie erforderlich ist, welche im allgemeinen Falle nicht gleich der Aktivierungsenergie der Selbstdiffusion im Volumen der Schmelze oder in der festen Phase sein wird.

Literatur

Akulov, N. S.: Dokl. AN **39** (7) (1943); predislovie b knige Prigožin, I.: Vvedenie b Termodinamiku neobratimych prozessov, IL 1960

Cahn, J. W., Hilliard, J. E.: J. chem. physics **28**, 258 (1958)

Derjagin, B. V., Karasev, V. V., Sorin, S. M.: Strojenie i fisičeskie svojstva veščestva b židkom sostojanii, Kiev 1954

GIBBS, J. W.: Termodinamičeskie raboti III. O ravnovesii geterogenych veščestv, Moskva 1950
HILLIG, U., TURNBULL, D.: Elementarnye prozessy rosta kristallov, IL 1959, S. 293
JANTSCH, O.: Z. Kristallogr., **108**, 185 (1956)
KOBEKO, P. P., KUVŠINSKIJ, E. V., ŠIŠKIN, N. J.: Žurn. Techn. Fiz. **14**, 10 (1944)
LAUDISE, R. A.: Hydrothermal Growth in: The Art and Science of Growing Crystals, New York **1963**, S. 231
LJUBOV, B. Ja.: Dokl. AN **72**, 273 (1950); Problemy metallovedenija i fisiki metallov **5**, 294 (1958)
MELEŠKO, L. O.: Sb. Mechanism i Kinetika Kristallisazii, Minsk **1964**, S. 113
MELEŠKO, L. O.: Kristall und Technik **3**, **3** (1968)
PANČENKOV, G. M.: Teorija vjaskosti židkosti, Moskva-Leningrad **1947**
ŠIŠKIN, N. I.: Žurn. Techn. Fiz. **26**, (7) (1956)
VOLMER, M., ADHIKAZI, G.: Z. Physik **35**, 170 (1925)

(Eingegangen am 1. März 1969, Übers. am 2. April 1969)

Anschrift des Verfassers:

L. O. MELEŠKO
Minsk 12
Tolbuchinastr. 13, Whng. 35
BSSR

Kristall und Technik | 5 | 1 | 1970 | 109–119

V. Mareček, J. Novák

Institute of Physics of Chechoslovakian Academy of Sciences, Prague

Effect of Crystallisation Conditions on the Shape and Composition of Ferroelectric Crystals

In the present paper the effect of crystallization conditions on the habit and chemical composition of some water-soluble ferroelectric crystals as TGS, TGFB, TGSe, KDP, RbDP, mixed LiTl-$LiNH_4$-tartrates and other tartrates is discussed. The influence of impurities or additives as well as different crystallisation conditions on growth and quality of crystals and additional treatment of samples for physical measurements are followed. Further, some established facts are mentioned concerning the chemical composition of perovskite type crystals of alkaline earth titanates prepared by crystallisation from molten potassium fluoride. The amount of impurities and additives is discussed. Many factors which affect the real crystallisation and which must be considered when evaluating the results of physical meausrements are reported.

In der vorliegenden Arbeit wird der Einfluß der Kristallisationsbedingungen auf den Habitus und die chemische Zusammensetzung einiger wasserlöslicher ferroelektrischer Kristalle wie TGS, TGFB, TGSe, KDP, RbDP, Mischkristalle von LiTl-$LiNH_4$ Tartraten und anderer Tartrate diskutiert. Der Einfluß von Verunreinigungen oder Zusätzen wie auch verschiedener Kristallisationsbedingungen auf das Wachstum und die Güte der Kristalle wird untersucht und eine zusätzliche Bearbeitung der Proben für physikalische Messungen beschrieben. Weiterhin werden einige begründete Sachverhalte hinsichtlich der chemischen Zusammensetzung von Erdalkalititanat-Kristallen des Perowskittyps (kristallisiert aus einer KF-Schmelze) angegeben. Die Menge an Verunreinigungen und Zusätzen wird besprochen. Eine Reihe von Faktoren, die das Realwachstum beeinflussen und bei der Auswertung physikalischer Messungen beachtet werden müssen, wird diskutiert.

1. Introduction

It is known that physical measurements performed on crystals of the same substance, but prepared in different laboratories, and even on samples from different parts of the same crystal, sometimes show different values. Also ageing of crystals provides different effects similarly as heat treatment does performed under comparable conditions but in different laboratories. An acceptable explanation for these phenomena has not yet been presented. An opinion still predominates that their origin must be sought in chemical impurities. We know from our experience that the same results occur when the crystallisation from the same raw material was carried out. In addition, cases are also known in which crystals prepared by crystallisation from water solution or melt, differ in physical or other properties. Also theories describing the crystallisation process start from an ideal case of purest material and from very simplified conditions of crystallisation. As it will be shown later, the real crystallisation process is much more complicated so that it is only little similar to the theoretical model or it only proceeds at short time intervals with apparent accordance. The next consideration will touch this complicated problem at least from one point of view.

2. Purity of raw materials and additives

Impurities present even as traces in the initial solution of the crystallizing substance will enter the solid phase according to the distribution coefficients of each impurity. Even if we know all distribution coefficients and the equilibrium diagram, we find that an effective distribution coefficient can attain another value during crystallisation, because it also depends on the growth rate of the crystal, on the diffusion coefficient and on the intensity of stirring. If we assume that the equilibrium distribution coefficient has a value lower than one, the separating effect will be greater during slow crystal growth and quick stirring than during rapid crystal growth and slow stirring. Providing the value of the equilibrium distribution coefficient is >1, an opposite situation arises, i. e. the separating effect increases with increasing rate of growth.

If during crystallisation the impurities accumulate in the mother liquor, then the growing crystal becomes continuously less pure and its final chemical composition will not be uniform, i.e. a higher content of impurities will be present on the surface than inside. If we admit the possibility of formation of mixed crystals (crystallizing substance and impurities), we can also take into consideration the coefficient of the Doerner-Hoskins equation, which depends, of course, on crystallisation conditions. Further, it is possible that there is among impurities present one which even accelerates or causes precipitation of another impurity. Similarly, we also could consider other factors, for example, diffusion, Coulomb forces, the influence and activity of solvent, *p*H of solution, equilibrium of electric charges, relative ionic radius, formation of aggregates and of mixed crystals, chemical affinity, solubility under different conditions, anisotropy, occlusions etc., whose action depends on the conditions of crystallisation.

When describing individual crystallisation theories we start from a conception of the growth of a chemically pure individuum from a solution of an exactly defined solvent. In practice, no absolutely pure compound or solution exists, and consequently, even after a most careful purification of substance we always deal with a mixture, i. e. the starting, purified compound is accompanied by impurities (even in traces). Although some theories of crystallisation with additives have been developed, the variety of results confirms our assumption of a very complicated crystal growth (MAREČEK; KOMENSKAYA; KAHLWEIT). The accompanying impurities are chemical substances with individual physical and chemical properties which under different conditions react differently (solubility, reactivity, magnetic or electric moment, density, behaviour in the proximity of the crystal face or the influence on the way of deposition in the crystal lattice, etc.).

In crystals prepared from flux thermal dissociation, partial pressure of oxygen, viscosity of the flux and diffusion in it, temperature gradient as well as properties of the furnace of given construction, and material of crucible may play an important role. An idea still predominates, that it is sufficient to perform crystallisation of chemically pure compound in any apparatus available and that the results will be the same. As a matter of fact, final properties of a crystal will depend on many accidental conditions. Analogical considerations apply also to artificial doping the initial crystallisation mixture.

3. Influence of crystallisation arrangement

Besides purity of starting materials, the own arrangement of crystallisation plays an important role. Let us consider, for example, the crystallisation of TGS. We can produce crystals from a solution either by a temperature, decrease or at constant temperature by gradual supersaturating the solution or by evaporating the solvent. In view of the defects, dislocations and other physical and chemical factors at a given temperature, it is evident that in the first case the mass of a crystal could by unhomogeneous, during crystallisation at decreasing temperature, and on the contrary, isothermal crystallisation offers some possibility to get more homogeneous substances (structure). The problem, however, is more complicated. In the first case already temperature decrease can be linear or non-linear, this being dependent on the surface of the grown crystal. Only in the former case a uniform growth is ensured. It is very difficult, however, to govern this condition technically (automatical temperature lowering according to the surface of the grown crystal). Even though in this direction a mathematical solution was achieved and expirimentally tested (JANTA, MASTNER, NOVOTNÝ).

The crystal grows either statically, when diffusion is the predominating factor of growth speed, or under dynamic conditions (various shapes of stirrers and different number of revolutions) under or without vibrations under the influence of ultrasonics, radiation, magnetic or electric field, in a spacious or small vessel, under constant temperature with evaporation of the solvent or melt or during continuous supersaturation at temperature decrease with different decrease rates from different starting values. The crystal grows in a horizontal or vertical direction (down or upwards) the growing crystal being supplied the supersaturated solution from one side, around or in the direction according to one axis, slowly or intensively, from a seed suitably chosen prepared and placed, in a certain way, from a solution prepared by different technology and adjusted (established) to a certain pH value, by repeatedly used solution and by different ways prepared for crystallisation. If we further consider that during a long-termed crystallisation certain factors affect the solution (changes in pH, colour, concentration and composition of impurities etc.), then the conception of growth of an inhomogeneous substance under different conditions of crystallisation is entirely inevitable. Just these factors are responsible for different results achieved during crystallisation of the same substance by different authors, even during repeated crystallisation in the same laboratory (see also KOLDOBSKAYA, MELESHINA, RES, MIRONOVA, GAVRILOVA, SAFRONOV).

4. Treatment of prepared crystals

Also further treatment and conditions in which the crystal is used for experimental purposes determine its properties. Many substances undergo changes caused by humidity, radiation or temperature treatment, especially those, which contain halogen (AREND, COUFOVÁ, NOVÁK) or hydroxyl (COUFOVÁ,

Table 1

Some watersoluble crystals, crystallisation conditions and crystal properties

Substance	Material	Impurities Additives	arrangement		growth direction
			Dynamic	Static	
TGS	origin.	0,01 % Fe, Pb, As, Si	revers. 20 – 40 r.p.m.	–	down
	origin.	ditto	ditto	–	down
	3 × cryst.	ditto, but traces	ditto	–	ditto
	ditto	ditto, with add. 0.1 % Tl	ditto	–	ditto
	ditto	ditto, with add. 0.1 % Cu	ditto	–	ditto
	origin.	0.01 % Fe, Pb, As, Si	–	without mix.	up
TGFB	orig. (met. Be)	Tr, Fe, Mg, Al, Sr, Si, Ca, Ba	revers. 20 – 40 r.p.m.	–	down
	2 × cryst.	ditto traces (weak)	ditto	–	down
	orig. (from BeO)	0.01 % Fe, Mg, Al, Sr, Si, Ca, Ba	ditto	–	down
	orig. (from met. Be)	Tr. Fe, Mg, Al, Sr, Si, Ca, Ba	300 r.p.m.	–	down
	2 × cryst. (from Be)	ditto traces (weak)	–	without mix.	up
	2 × cryst. (from Be)	ditto, addit. 0.1 % Fe	20 – 40 r.p.m.	–	down
	2 × cryst. (from Be)	ditto, addit. 0.1 % Cu	ditto	–	down
	2 × cryst. (from Be)	ditto, addit. 0.1 % Ni	ditto	–	down
	2 × cryst. (from Be)	ditto, addit. 0.1 % Cr^{3+}	ditto	–	down
KDP	p. a.	Tr. Fe, Al, Si, Ba, Sr, Mg, Cu	revers. 20 – 40 r.p.m.	–	up
	p. a.	ditto	ditto	–	up
	p. a.	ditto	ditto	–	down
	p. a.	ditto	ditto	–	horizontal
	p. a.	ditto	ditto	–	up
	p. a.	ditto	ditto	–	up
	p. a.	ditto	ditto	–	up
	p. a.	ditto	ditto	–	up
	p. a.	ditto a. 0.1 % Fe, Ni, Sn, Cr, Al	–	without mix.	up
	p. a.	ditto ditto	revers.	–	up
	1 × cryst.	ditto fewer		–	down
RbDP	p. a.	Tr. Fe, Mg, Si, Ca, Ag, Bi, As	revers. 20 r.p.m.	–	up
	p. a.	ditto	–	without mix.	up
	p. a.	ditto	revers. 20 r.p.m.	–	down
LiTl tartrate	orig. 0.1 % Al, Cu, Fe, Mg, Si, Ca, Ag, Bi		revers. 40 r.p.m.	–	down
	ditto	ditto	–	without mix.	up
	2 × cryst.	ditto in traces	revers. 40 r.p.m. vibration	–	down
	ditto	ditto in traces	–	without mix.	up
	orig. 0.1 % Al ect.		revers. 40 r.p.m. vibration	–	up
	ditto rep. cryst.	of the same solution	–	without mix.	up
	ditto II. rep. cryst.	of the same solution	–	without mix.	up
	2 × cryst.	ditto in traces	revers. 300 r.p.m.	–	down
	ditto	ditto in traces	revers. 20 r.p.m.	–	up
	orig. 0.1 % Al etc.		revers. 20 r.p.m.	–	down
	2 × cryst.	ditto in traces	revers. 20 r.p.m.	–	down
	2 × cryst.	ditto in traces	revers. 20 r.p.m.	–	down
LiAm tartrate	orig. 0.1 % Al, Mg, Ca, Pb, Si		revers. 20 r.p.m.	–	down
	ditto	ditto in traces	revers. 20 r.p.m. vibration	–	down
	ditto	ditto in traces	–	without mix.	up
KK tartrate	p. a. traces Ca, Mg, Si, Cu, Al		revers. 20 r.p.m.	–	down
	ditto		revers. 20 r.p.m.	–	down
EDT	p. a. traces Ca, Mg, Si, Cu, Al		revers. 20 r.p.m.	–	down
	p. a. ditto		revers. 20 r.p.m.	–	down
Sol. 10 % – 90 % LiAm	orig. 0.1 % Al, Mg, Ca, Pb, Cu, Fe, Si, Ag, Bi		revers. 40 r.p.m.		down
90 – 10 % LiTl tartrates	ditto ditto		vibration –	without mix.	ditto

seed	pH solution	Temperature	Temperature change	Properties of the crystal	Impurities
prism ⊥ b axis	1.5	45 °C decreas.	slow	many faces	tr. Fe, Pb, Si, As
ditto	1.5	ditto	fast	few faces	ditto
ditto	4.2	ditto	fast	few faces	weak tr. ,,
ditto	4.2	ditto	middle	no growth in c	ax. ditto, tr. Tl
ditto	4.2	ditto	middle	perfect cryst.	ditto, Cu proved
ditto	1.5	26 °C	–	normal growth	tr. Fe, Si, Pb
prism ⊥ b axis	1.5	46 °C decreas.	slow	one face def.	tr. of all
ditto	2.5	46 °C ditto	slow	perfect cryst.	weak tr.
ditto	1.5	46 °C ditto	slow	defects	all proved
ditto	1.5	ditto	slow	defects	tr. of all
ditto	2.5	30 °C	–	perfect	ditto
ditto	2.5	30 °C evap.	–	good cryst.	Fe and tr. of all
ditto	2.5	ditto	–	perf. cryst. blue	Cu and tr.
ditto	2.5	ditto	–	ditto, green	Ni and tr.
ditto	2.5	ditto	–	many defects	Cr and tr.
plate ⊥ c axis	3.5 – 4.2	40 °C decreas.	slow	wodged	all proved
ditto	ditto	80 °C ditto	slow	wedged	all proved
ditto	ditto	40 °C ditto	slow	wedged (fewer ?)	ditto
ditto	ditto	40 °C ditto	slow	wegded (fewer ?)	ditto
ditto	4,8	40 °C ditto	slow	few wedged	all pres.
ditto	5.0 – 5.5	40 °C ditto	slow	non-wedged	all pres.
ditto	6.0 – 6.6	40 °C ditto	slow	ditto growth in a, b	axis ditto
ditto	5.0 – 5.5	40 °C ditto	fast	non-wedged	all pr. fewer ?
ditto	5.1	26 °C –	–	non-wedged	only Fe traces
ditto	5.3	40 °C decreas.	slow	non-wedged, incrust.	non-homog.
plate from cr. pH 4.1	5.3	40 °C decreas.	slow	ditto, cracks	all present
plate ⊥ c axis	5.25	40 °C decreas.	slow	non-wedged, normal	tr. of all
ditto	5.3	26 °C	–	non-wedged, perfect	few. tr.
ditto	3.6 – 4.2	40 °C decreas.	slow	wedged, defects	all present
1. par. cryst.	7.0	40 °C decreas.	slow	defect. faces	all present
ditto	7.0	40 °C	–	normal	ditto fewer
ditto	7.0	34 °C decreas.	slow	some def. faces	ditto fewer
ditto	7.0	34 °C	–	perfect	ditto fewer
ditto	7.0	40 °C decreas.	slow	face ag. mix. defect	tr. of all
ditto	7.0	26 °C	–	hab. normal, greenish	ditto
ditto	7.0	26 °C	–	def. cryst. green.	ditto, all
ditto	7,0	41 °C decreas.	slow	defect cryst.	tr. of all
prism. of 010	7.0	40 °C decreas.	slow	normal	tr. of all
ditto	7.0	41 °C decreas.	slow	after chang. pos.	growth of def. f.
ditto	7.0	38 °C decreas.	slow	normal cryst.	tr. of all
ditto	7.0	38 °C decreas.	fast	normal cryst.	tr. of all
1. apr. cryst.	7.0	40 °C decreas.	slow	normal cryst.	tr. of all
ditto	7.0	40 °C decreas.	slow	def. crystal	ditto
ditto	7.0	26 °C	–	normal perfect	ditto
par. cryst.	11.0	40 °C decreas.	slow	normal	tr. of all
ditto	11.0	40 °C decreas.	slow	defect	tr. of all
part. of cryst.	8.0	50 °C decreas.	slow	normal	–
ditto	8.0	35 °C decreas.	slow	defects	–
par. cryst.	7.0	39 °C decress.	slow	ca. 35 % LiAm	tartrate, defects
ditto	7.0	26 °C		ca. 90 % LiAm see Tab. 2	tartrate, normal

Novák, Hlasivcová) in the crystal lattice as defect. We are convinced that the only right way is to consider properties of crystals as a whole complex of all affecting factors, including origin of raw materials, way of preparation, ageing, and temperature treatment. Only just after specifying all these data, a comparison of the results of measurements of the same parameters on the same material may be carried out.

The investigation of the influence of crystallisation method on habit, structure, and composition of prepared crystals leads to a conclusion that the preparation of an imperfect crystal ("Realkristall") is a very complicated process, the course of which as a whole, is not in accordance with the known theories of the crystal growth, since these theories consider very simplified processes and do not take into account a variety of factors which do not act spontaneously, but according to the experimental conditions thus providing different properties of prepared crystals of the same starting material.

5. Results

Tables are presented summarizing the results of growth of crystals and chemical composition of some water-soluble materials and some perovskite type materials grown from flux, the crystals being obtained under various conditions.

The results achieved during preparation of some water-soluble materials are shown in Table 1. More detailed data are summarized in Tables 2–4. Some results for perovskite materials are presented in Tables 5–8.

Figure 1 shows that in growing TGS crystals the number of crystal faces decreases with increasing speed of the temperature decrease. The presence of Tl^{I}ions negatively affects the growth along the c-axis. The TGS crystal grows

Table 2

Influence of crystallisation conditions on the composition of mixed LiTl-LiAm tartrates

Composition of mixture in %mol of tartrate LiTl: LiAm	Way of crystallisation	Temperature	LiAm tartrate content in the crystal [W%]
1:1.5	isothermal static	24.8 °C	14.7
1:5	ditto	24.8 °C	67.4
1:7	ditto	24.8 °C	83.6
1:8	ditto	24.0 °C	78.2
1:5	ditto	24.0 °C	52.4
1:2.5	ditto	24.0 °C	20.9
1:9	static with temperature decrease	39.5–32 °C	84.5
1:7	ditto	39.5–32 °C	70.4
1:1.5	ditto	39.5–32 °C	20.9
1:9	dynamic with temperature decrease	39.5–32 °C	35.7
1:7	ditto	39.5–32 °C	34.3
1:1.5	ditto	39.5–32 °C	13.5
1:1.5	isothermal dynamic	24.8 °C	12.5

Table 3
Growth of some tartrates during different temperature decrease.
Length in direction of different axis (mm)

Tartrate	Temperature decrease from 40 °C to 26 °C in 5 hours			Temperature decrease from 40 °C to 33 °C in 24 days		
	a:	b:	c:	a	b	c
Sodium Thallium	7.0 — 8.1	5.1— 7.7	5.1 — 7.6	2.8— 9.1	1.4—6.6	3.6—3.8
Normal Potassium	3.4 — 4.2	2.6— 3.8	2.9 — 3.7	3.4— 7.0	2.4—5.8	1.8—2.0
Lithium Thallium	15.25—17.4	11. —15.25	11.65—13.2	3.0—11.0	1.5—8.0	4.5—4.9

Table 4

Fe content (w %) in KDP crystals grown under different conditions. Raw material contains $2 \cdot 10^{-3}$ w% Fe

Way of crystallisation	(w %)
Quick temperature decrease	$5 \cdot 10^{-3}$
Slow temperature decrease	$1.4 \cdot 10^{-3}$
Isothermal, quiet	$1.08 \cdot 10^{-3}$

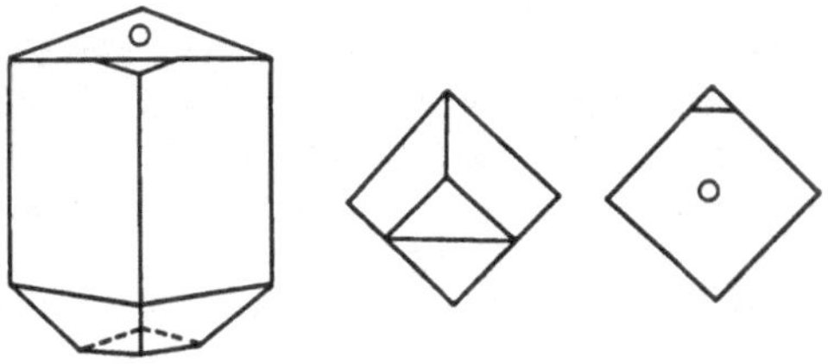

a

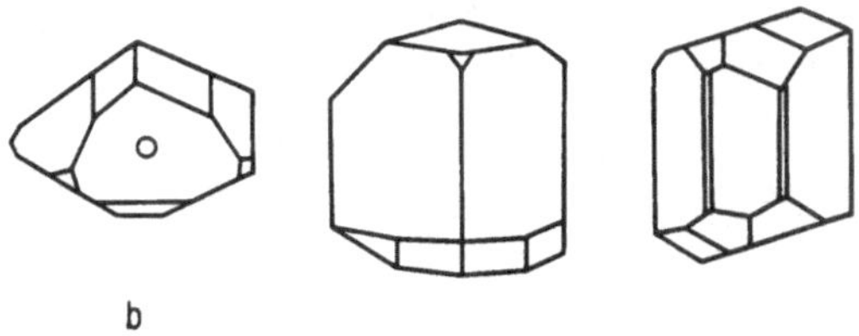

b

Fig. 1. Change of TGS crystal habit due to different rate of temperature decrease during crystallisation

a — quick cooling
b — slow cooling

at pH 1.5—4.2. The presence of additives, for example, Fe^{III}, Cu^{II} or Ni^{II} manifests itself to a different extent. While ferric and cupric ions are incorporated in the crystal lattice, nickel could not be proved in it. The growth in the static arrangement at a constant temperature is almost perfect. The regular growth of TGFB crystals is achieved by keeping purity of the raw material on a certain level. Increased content of impurities provides a face defect. Iron, copper, or nickel ions favour the regular growth and enter the crystal body, whereas Cr^{III} ions have a negative effect on the growth. It is possible to replace as much as 0.1 w% Be^{II} by Fe^{III}, and the growth remains unchanged. Stirring by a number of rotations higher than 300 r.p.m. disturbs the normal growth in the pH range 1,5—2.5. The potassium dihydrogen phosphate crystals grow at pH 5.0—5.5 as non-wedged crystals. More detailed information is presented by MAREČEK, DOBIÁŠOVÁ, NOVÁK. Pure raw material can only provide regular crystals of tartrates. If some impurities are present in the solutions, nearly perfect LiTl tartrate crystals can be obtained in the static arrangement only. Rapid stirring and vibration of the crystallisator take a negative influence on the growth, which can be affected by the seed, speed of temperature decrease and by stopping the stirring. By changing the position of the growing crystal during growth, more perfect crystal faces can be achieved. The pH value of the solution changes with time as well as its colour. Repeated crystallisations in the static arrangement lead to crystals of still deteriorating properties. Other tartrates can be succesfully prepared only under certain conditions (KK, ED, $LiNH_4$).

Table 3. shows that the composition of mixed tartrate crystals depends on the crystallisation method. The results on the influence of the speed of temperature decrease upon the differing growth of NaTl-, KK-, LiTl tartrates are presented in Table 3. In Table 4 analytical data of impurities present in the KDP crystals prepared from the same raw material by using the same crystallisation method with different crystallisation arrangements are given.

Similar effects were observed on water-insoluble crystals prepared from a mixture of molten salts at high temperatures. Tables 5—8 inform us that crystals of pure barium metatitanate (REMEIKA) obtained from the same charge are not of equal composition. The problem is complicated because of additives which are always present in the crystals either coming from the melt and from the crucible or being intentionally added to the initial mixture. Table 7 shows the large effect on the composition of pure $BaTiO_3$ crystals, if the amount of

Table 5

Fluorine and potassium contents in $BaTiO_3$ crystals prepared from molten KF at different temperature decrease during crystallization

Nr. of Experiment	Temperature decrease in °C/hour	Found (w %)	
		K	F
1	5	0.19	0.0996
2	10	0.21	0.095
3	20	0.22	0.11
4	60	0.35	0.168
5	120	0.308	0.139

Table 6

Dependence of fluorine content in $BaTiO_3$ crystals on the thickness of the crystal during constant temperature decrease

Thickness (μ)	Det. weight % F	K
30	0.40	0.54
50	0.29	0.35
100	0.12	0.25
150	0.087	0.20
180	0.033	0.12

Table 7

Analytical results achieved on $BaTiO_3$ crystals with different $BaO:TiO_2$ ratio crystallized according to REMEIKA method in a platinum crucible (K, F, Fe, Pt as impurities, not as additives)

Initial $BaO:TiO_2$ ratio of samples	BaO	TiO_2	K	F	Fe	Pt
	(w%)					
1.5 : 1	66.21	33.85	0.025	0.03	0.002	0.33
1.2 : 1	65.84	34.17	0.025	0.032	0.0043	0.148
1.07 : 1	—			0.08		0.176
1 : 1	64.8	34.26	0.22	0.11	0.005	0.056
1 : 1.01			0.22	0.11		0.047
1 : 1.07			0.43	0.155	0.005	0.035
1 : 1.2	63.8	34.62	0.58	0.0208	0.006	0.002
1 : 1.5	62.0	35.18	1.05	0.33	0.028	0.001

Table 8

Dependence of the composition of (Ba, Sr) TiO_3 crystals on the composition of initial mixture

Composition of initial mixture	Det. weight %		Det. Mol%	
	BaO	SrO	BaO	SrO
$BaO_{0.8}$ $SrO_{0.2}$ TiO_3	51.04	8.83	76.5	22.7
$BaO_{0.5}$ $SrO_{0.5}$ TiO_3	21.82	32.40	36.7	64.2
$BaO_{0.3}$ $SrO_{0.7}$ TiO_3	13.93	42.20	17.5	83.0
$BaO_{0.1}$ $SrO_{0.9}$ TiO_3	3.24	53.32	4.92	94.5
$BaTiO_3$	64.8	0.0	98.2	0.0
$SrTiO_3$	0.0	56.22	0.0	99.6

one of the fundamental components of starting material (in this case ratio $BaO:TiO_2$) is changed. This results are essentially influenced not only by the change of potassium and fluorine contents, but also of platinum percentage and stoichiometry of resulting oxides. Table 5 shows the influence of speed of temperature decrease on fluorine and potassium contents in barium titanate single crystals. Table 6 gives analytical data concerning the fluorine content in dependence on crystal thickness. Table 8 demonstrates the essential influence of isovalent substitution of strontium for barium in barium titanate crystals according to the BaO:SrO ratio.

The presented examples show that the preparation of perfect crystals from solutions or molten salts is influenced by many factors which vary with individual crystals. The influence of any factor is very difficult to predict and for that reason it is always necessary to establish it experimentally, because none of the existing theories (see below) is fully capable to explain all conditions as well as mechanism of crystallisation. Considerations on the hydrodynamic properties of solutions, and the requirement for a most efficient contact of the growing crystal with the whole liquid volume concern the crystallisation arrangement, stirring, shape of stirrer and number of rotations and physical properties of the solution, but they are rather general and in the best case they only take into account crystal size (Kasatkin; Lengyel; Nývlt (1)). It is true that the whole theory of nucleation deals with the influence of temperature, explains the conception of fluctuation and concentration of the solution, describes the mutual influence of particles in a real solution, but all these considerations are general and for the own mechanism of the crystallisation do not bring detailed informations for the given crystal growth (Howard; Nývlt (2)). Papers dealing with the metastabile region of the solution are, indeed, very valuable but for the explanation of different crystallisations they do not give a statisfactory reply (Todhunter). Detailed studies on mixing (Kubo, Tanaka; Garrett, Rosenbaum) by supersonics (Kapustin, Kavaluneite), and the magnetic field (Mayr) give explanations on properties of prepared crystals (chiefly with respect to size and cohesion of crystals), but they are not sufficient for our purposes. Interesting considerations about additives (Gupta, Buckley) chiefly deal with crystal size and formation of soluble binary compounds in connection with the surface energy of crystals.

The mentioned examples represent the beginning of a detailed further study which will try to elucidate crystal growth under conditions of the real crystallisation.

The authors wish to express their thanks to Dr. P. Coufová and Ing. J. Hejduk for reading the manuscript and for valuable comments.

References

Arend, H., Coufová, P., Novák, J.: J. Amer. chem. Soc. **50**, 22 (1967)
Coufová, P., Novák, J., Hlasivcová, N.: J. Chim. physique **45**, 3171 (1966)
Garrett, D. E., Rosenbaum, G. P.: Chem. Engng. **65**, 125 (1958)
Gupta, I. R. K., Buckley, H. E.: Z. Physik **204**, 165 (1955)
Howard, J. R.: Trans. Faraday Soc. **56**, 278 (1960)
Janta, J., Mastner, J., Novotný, J.: II. celostátní seminář o monokrystalech, Hrubá skála-Turnov, duben 1967
Kahlweit, M.: Z. physik. Chem. **25**, 125 (1960)

KAPUSTIN, A. P., KAVALUNEJTE, V. E.: Sb. primenenija ultrazvuka **9**, 127 (1959)
KASATKIN, A. G.: Osnovnye processy i apparaty chimičeskoj technologii. Goschimizdat, Moskva 1948
KOLDOBSKAYA, M. F., MELESHINA, V. A., RES, I. S., MIRONOVA, S. A., GAVRILOVA, I. V., SAFRONOV, G. M.: Kristallografiya **11**, 785 (1966)
KOMENSKAYA, I.: Rost kristalov, **39** (1957)
KUBO, T., TANAKA, K.: Kogyo Kagaku Zasshi **59**, 273 (1956)
LENGYEL, S.: Termodinamika i strojenie rastvorov, Trudy i sověčš. AN SSSR 144, (1958)
MAREČEK, V.: Czech. J. Phys. A **19**, 121 (1969)
MAREČEK, V., DOBIÁŠOVÁ, L., NOVÁK, J.: Kristall u. Technik **4**, 39 (1969)
MAYR, G.: Z. Naturforsch. **A 6**, 467 (1951)
NÝVLT, J.: Chem. Průmysl **12**, 440 (1962) (1)
NÝVLT, J.: Chem. Průmysl **12**, 63 (1962) (2)
REMEIKA, J. P.: J. Amer. chem. Soc. **76**, 940 (1954)
TODHUNTER, J.: Chem. and Ind. **29**, 1399 (1955)
USP 2 876 083 (1959)

(Received April 15, 1969)

Authors' address:

V. MAREČEK
Dr. J. NOVÁK
Institute of Physics of Czech. Acad. Sci.
Praha 2, Viničná ul 7

| Kristall und Technik | 5 | 1 | 1970 | 121–127 |

G. Bliznakov, S. Yaneva

Department of Chemistry of the University, Sofia

Effect of Adsorption on the Dissolution of Crystals (I)

Effect of Phenol on the Rate of KBr Dissolution

Kinetic investigations of the effect of phenol on the rate of dissolution of KBr cubic faces have been made. The heat of adsorption (λ_a) on the active sites of dissolution has been determined. This adsorption heat is found to be larger than the heat of adsorption on the active sites of growth previously determined for the same system, and three times larger than the adsorption heat of phenol on the whole surface of KBr cubic faces.

Были совершены кинетические исследования влияния фенола на скорость растворения кубических стен кристаллов KBr. Найдена величина адсорбционной теплоты фенола на активные места растворения. Оказалось что эта теплота больше теплоты адсорбции фенола на активные места роста кристаллов KBr найденая раньше и в три раза больше адсорбционной теплоты фенола на общую поверхность кубических стен.

1. Introduction

In many cases, adsorption of impurities causes considerable changes in the kinetics of growth of different crystal faces and hence, in the crystal habit (Buckley; Bliznakov 1957; Bliznakov, Kirkova; Bliznakov 1958, 1965). The influence of adsorption on the reverse process, i.e. the dissolution of crystals, has been studied to a smaller extent. The first investigations in this field are those of Steinike and Neels; Steinike who studied the effect of ferrohexacyanides on the dissolution of NaCl. In a number of later papers (Mutaftschiev, Chajes, Gindt; Van Damme-Van Weele; Nano, Miura), the same problem is treated using various systems. The prevailing opinion is that the role of impurities consists in poisoning parts of the surface owing to which the rate of dissolution decreases. Most of the above cited authors treat the problem qualitatively or semiquantitatively. Since crystal growth and dissolution have not been studied under similar conditions in one and the same system, there are no data concerning the extent to which the effect of impurity on crystal growth is symmetrical to its effect on dissolution. The purpose of the present work was to elucidate this problem by studying the kinetic effect of phenol on the rate of dissolution of cubic faces of KBr. Since the crystal growth of KBr in the presence of phenol had been investigated in detail in a previous work (Bliznakov, Kirkova, Nikolaeva), this enabled us to compare the results obtained in both cases.

2. Experimental

Experiments were carried out in a circulation apparatus employed in previous investigations (BLIZNAKOV, KIRKOVA; BLIZNAKOV 1965). The final model of the apparatus (after some improvements directed towards a better stabilization of temperature, especially in the left part of the apparatus and leading to a considerable diminution of its size) is shown in Figure 1.

Potassium bromide recrystallized with bidistilled water and freshly distilled phenol were used for the experiments. The KBr solutions were prepared using bidistilled water too. Measurements were made on well formed small crystals of cubic habit and dimensions of 1 – 2 mm, obtained by slow crystallization.

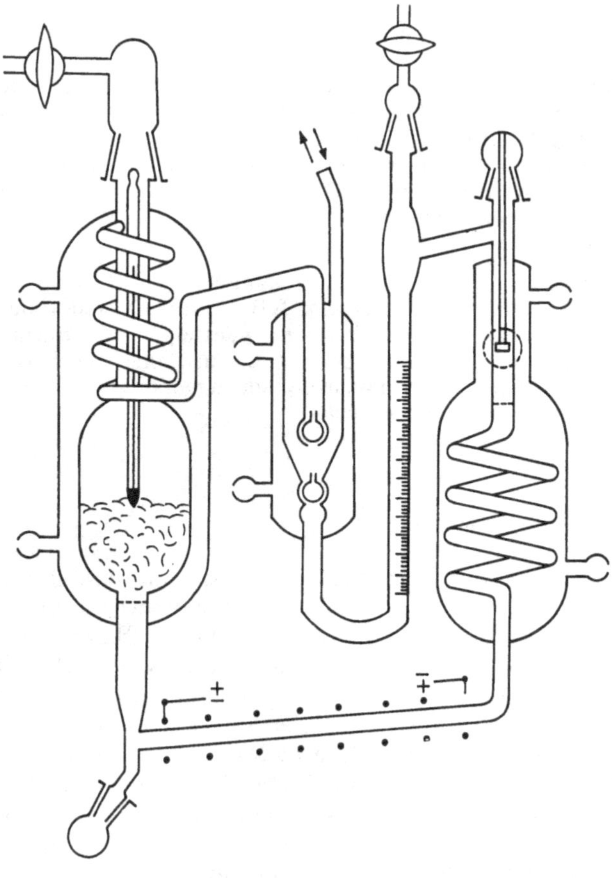

Fig. 1. Scheme of apparatus used for measuring crystal dissolution rates

The undersaturation was determined according to the solubility curve. Thermodynamical undersaturation was used:

$$\Delta\mu = R\,T \ln\,(C_s^{\mathrm{I}}/C_s^{\mathrm{II}})\,.$$

Therein C_s^{I} is the concentration of the saturated solution at temperature T_1 which is maintained in the left section of the apparatus; C_s^{II} is the concentration corresponding to the saturation at the temperature T_2 maintained in the right section where dissolution takes place.

Since $T_2 > T_1$, it follows that $C_s^{II} > C_s^{I}$. A sufficient accuracy was obtained using $(C_s^{II} - C_s^{I})/C_s^{II} = \Delta C/C_s^{II}$.

Experiments were performed at undersaturations ranging between 0,9 and 2,5%. The fact that the rate of dissolution is measurable in this range shows that the surface of crystals used is rich in sites of dislocation origin where the destruction of lattice layers begins. As was to be expected, corners and edges dissolved faster than other parts and after a certain time cubic crystals acquired an almost spheric symmetry (Figs. 2a, 2b, 2c). For this reason the rate of dissolution was measured in the middle of the faces during a period of 5—10 min in which the rate remained practically constant (Fig. 3) and the crystal habit showed almost no change. (One division on the ordinate corresponds to 2,84 · 10^{-4} cm).

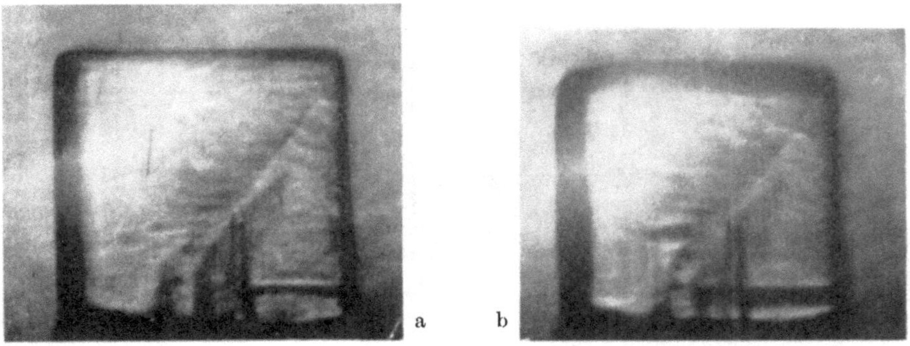

a b

c

Fig. 2. Crystal of KBr in the course of dissolution:
a – at the beginning of the process
b – 3 min later
c – 12 min later

In order to eliminate diffusion effects, it was necessary to study the influence of the rate of flow on the rate of dissolution. The results are shown in Figure 4. The rate of flow (in cm min^{-1}) is marked off along the abscissa. The curves plotted are analogous to those obtained by Coulson and Richardson in the case of crystal growth. A difference is observed at relatively low undersaturations, at which our curves are S-shaped. An essential point in this case is that the working conditions permit the realization of dissolution independent on the rate of flow. It was further shown that the extreme value of the rate of flow still affecting the rate of dissolution is practically independent of temperature when the supersaturation remains constant. This makes the choice of working conditions considerably easier. The rate of dissolution was measured at an undersaturation of 2,4%, a rate of flow of 120 cm min^{-1} and temperatures of dissolution of 16, 19, 21, 23 and 26 °C. Each point of the curves is a result from 5—6 measurements. The rate of dissolution was measured in divisions corresponding to $2{,}84 \cdot 10^{-4}$ cm.

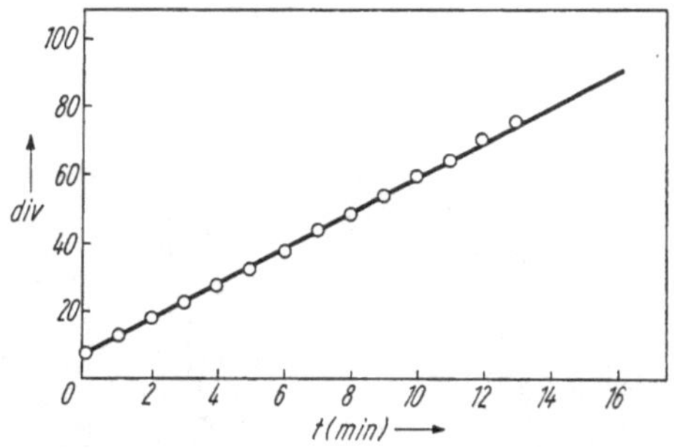

Fig. 3. Shift of the boundary crystal/dissolution versus time of dissolution for the face (100)

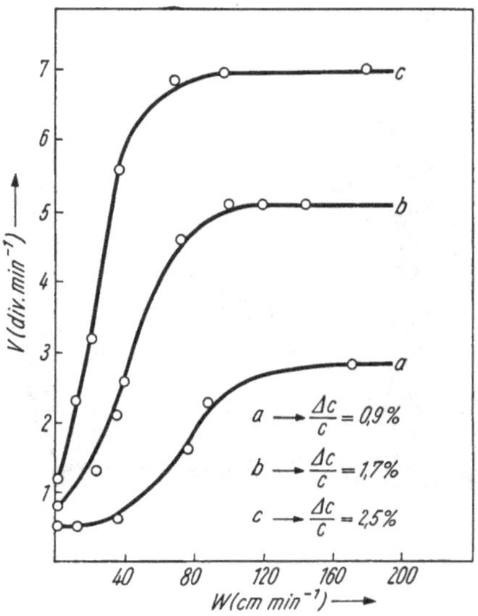

Fig. 4. Dependence of the rate of dissolution of a KBr crystal on the linear velocity of the solution

3. Results and discussion

The isotherms in Figure 5 show the changes in the rate of dissolution of KBr crystals as a function of the phenol concentration for five different temperatures. It is evident that the rate of dissolution decreases with the increase of impurity concentration, i.e. the effect is analogous to that observed with crystal growth (Bliznakov, Kirkova, Nikolaeva). This suggests an analogous interpretation of the results.

A phenomenological theory explaining the effect of impurity adsorption on crystal growth was developed in previous investigations (Bliznakov 1954, 1958, 1965) by assuming that a sufficient number of active centers of crystal growth or dissolution exist on the crystal surface due to various reasons. These centers may be blocked by adsorption of impurity particles (molecules or ions), which causes a formal division of the substance flowing to or from the crystal surface into two parts: one passing over the free and the other over the occupied surface. In the extreme case the flow over the occupied part of the surface may be equal to zero. This is the case of relatively great forces of adsorption existing between the impurity particles and the active sites of growth or dissolution.

For the rate of growth V, the following equation was derived:

$$V = V_0 - (V_c - V_\infty)\,\theta(c) \tag{1}$$

or

$$V = V_0\,[1 - \alpha\,\theta(c)] \tag{2}$$

Here V_0 is the rate of growth in pure solutions, V_∞, the rate of growth at maximum coverage of the active sites with adsorbed impurity particles ($V_0 \geqq V \geqq 0$), and $\theta(c)$ is the coverage, which is a function of concentration. Evidently $\alpha = 1 - (V_\infty/V_0)$. By introduction of the dimensionless quantities $\eta = V/V_0$ and $\eta_\infty = V_\infty/V_0$ as well as of a Langmuir adsorption isotherm, from (1) we obtain:

$$\frac{1}{1-\eta} = \frac{B}{1-\eta_\infty} \cdot \frac{1}{c} + \frac{1}{1-\eta_\infty} \tag{3}$$

where c is the impurity concentration and B the Langmuir constant.

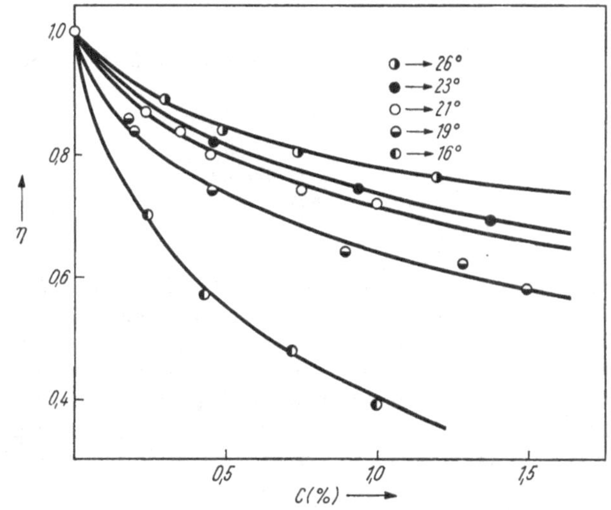

Fig. 5. Relative decrease of the rate of dissolution in dependence on impurity concentration in the solution at different temperatures

Equation (3) has proved to be valid in many cases, which supports the above conception on the mechanism of crystal growth in the presence of impurities (BLIZNAKOV 1958 1965). It could be expected that this conception on crystal growth would be valid for dissolution of crystals as well.

The experimental data treated according to equation (3) are presented in Figure 6. All isotherms show a satisfactory linear relationship. The Langmuir constant B is calculated for each temperature from the slope of the corresponding most probable curve. These data are represented on Figure 7 according to the equation

$$\lg B = -\lambda_a/2{,}3\ R\ T + \text{const}\ , \tag{4}$$

which gives the possibility of calculating the heat of adsorption λ_a on the active sites of dissolution. The value obtained from the slope of the most probable curve is $\lambda_a = 7800$ cal mol^{-1}. This value is higher than the values obtained previously (BLIZNAKOV, KIRKOVA, NIKOLAEVA) for the heat of adsorption of phenol on the active sites of the cubic faces of KBr. It is a point of interest that the adsorption heat so obtained is exactly three times greater than the heat of adsorption of phenol on the whole surface of the cubic faces. The value

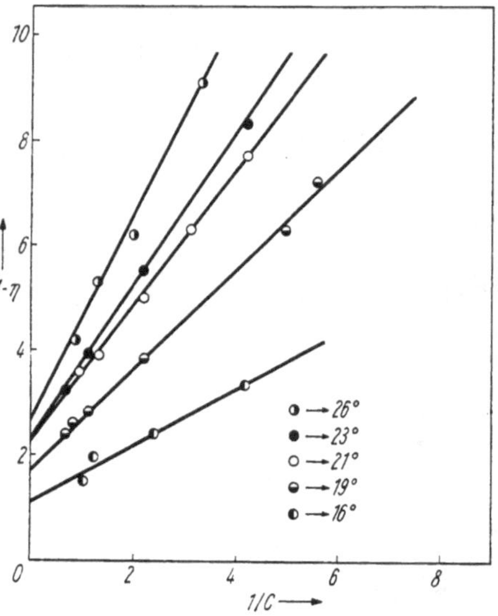

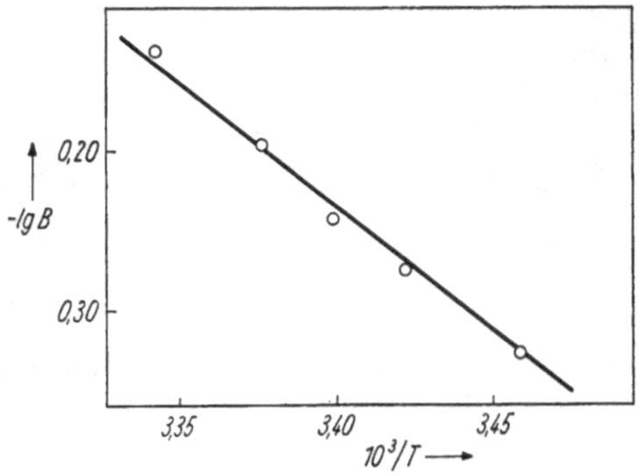

Fig. 7. Dependence of lg B on the reciprocal absolute temperature

Fig. 6. Dependence of $1/1-\eta$ on the reciprocal concentration at different temperatures

of the latter, determined previously (BLIZNAKOV, KIRKOVA, NIKOLAEVA), is $\lambda_a = 2600$ cal mol^{-1}. This coincidence, if not a chance one, supports the model of a molecule adsorbed by three adsorption valences (KNACKE, STRANSKI). Ideal as this model is, it is very convenient for the interpretation of adsorption processes in the case under consideration as well as in other analogous cases. According to this model, the adsorbed molecule is bound to the plane face by one of its valences, while three valences are engaged in the half crystal position, the heat of adsorption being three times larger in this case.

The question arises why the heat of adsorption of the same molecules on the same crystals appears to be lower ($\lambda_a = 5100$ cal mol^{-1}) when determined in growth conditions. Most probably it is a matter of correspondence in size between the molecule of phenol and the thickness of growing or dissolving layers. The layers of crystal growth are supposed to be thinner than the layers of dissolution. Perhaps the molecule of phenol, being very large, does not fit well in the half crystal position in the case of growth, so that its interaction with the neighbouring ions is hindered. The dipole-ion interaction appears to be stronger in the case of dissolution, due to the more coarse layer structure.

The present investigation shows that in the case studied the effect of impurity on dissolution is symmetrical to its effect on crystal growth, this symmetry reaching to the details of the processes.

The authors wish to express their thanks to Mrs. Emilia Metchenova for her participation in the performance of the experiments.

References

BLIZNAKOV, G.: Izv. Bulg. Akad. Nauk., otd. fiz. mat. nauki **4**, 135 (1954)

BLIZNAKOV, G.: Crystallografiya **4**, 150 (1957)

BLIZNAKOV, G.: Fortschr. Mineralogie **36**, 149 (1958)

BLIZNAKOV, G.: Adsorption et croissance cristalline — Edition de C.N.R.S, Paris 1965

BLIZNAKOV, G., KIRKOVA, E.: Z. physik. Chem. **206**, 271 (1957)
BLIZNAKOV, G., KIRKOVA, E., NIKOLAEVA, R.: Z. physik. Chem. **228**, 23 (1965)
BUCKLEY, H. E.: Crystal growth, London 1951
COULSON, Y. M., RICHARDSON, J. F.: Chemical Engineering, Vol. 2, 817, London 1955
VAN DAMME-VAN WEELE, M. A.: Adsorption et croissance cristalline — Edition de C.N.R.S, Paris 1965
KNACKE, O., STRANSKI, I.: Z. Elektrochem. Ber. Bunsenges. physik. Chem. **60**, 816 (1956)
MUTAFTSCHIEV, B., CHAJES, H., GINDT, K.: Adsorption et croissance cristalline — Edition de C.N.R.S, Paris 1965
NAONO, H., MIURA, M.: J. Sci. Hiroshima Univ. ser. A Div. 2, **29** Nr. 3, 81 (1965)
NEELS, H., STEINIKE, U.: Freiberger Forsch. H. A **267**, 433 (1963)
STEINIKE, U.: Z. anorg. allg. Chem. **317**, 186 (1962)

(Received April 23, 1969)

Authors' address:

Professor Dr. G. BLIZNAKOV
Mrs. S. YANEVA
Institut für anorganische und allgemeine Chemie der Universität Sofia
Sofia/Bulgarien
Boulevard Anton Ivanov Nr. 1

Kristall und Technik | 5 | 1 | 1970 | 129–144

V. A. FRANK-KAMENETZKIJ, N. V. KOTOV, E. A. GOILO,
V. A. SCHITOV

Lehrstuhl für Kristallographie und Institut für Erdkrustenforschung
der Staatlichen Universität Leningrad

Strukturveränderungen von Tonmineralien unter verschiedenen thermodynamischen Bedingungen (III)

Transformative Umwandlungen von Mineralen der Kaolinit-Gruppe unter erhöhtem Druck bei Anwesenheit von mineralisierenden Lösungen [1])

In the earlier papers (I and II, Frank-Kamenetzkij et al., 1966, 1968) data regarding the structural and facial changes in kaolin group minerals under conditions of interaction with hydrothermal chloride solutions of K, Na, Ca, Mg at $P_{H_2O} = 900$ kg cm^{-2}, $T = 200-500$ °C were dis- cussed. In the present paper the results of these investigations are summed up.

The diffractometrical, electronographic and electron microscopic investigations of initial, intermedial and final phasis mainly of layered silicate and aluminous silicates show a cristallographic inheritance of structures during the transformation of kaolinite into montmorillonite, micas and mixed layered facies (rectorit, tosudit etc.) This is seen in the retention of the main features in the basal plane, in the character of population of the octahedral positions, in the tendency to retain the sequence of alternation and imposition of layers (polytype), and in the processes of deformation that the structures of the initial minerals undergo during such changes. The wide participation of ordered and disordered mixed-layered structures in these reactions is the result of structural changes by transformatical means. At the same time, the chemical nature of the medium exerts considerable influence on the course and nature of the resulting structures. Thus, the mechanism of transformational thermal changes of aluminium silicates, which is already known for "dry-systems", is observed to a great extent in the case of complex polycomponental systems under H_2O pressure and is characterised by the specific individual features discussed in this paper.

В предыдущих сообщениях (I и II. Frank-Kamenetzkij u. a., 1966, 1968) был приведен фактический материал о структурных и фазовых преобразованиях каолиновых минералов в условиях взаимодействия с гидротермальными растворами хлоридов K, Na, Ca, Mg при $P_{H_2O} = 900$ кГ см$^{-2}$, T = 200–500 °C. В настоящем, 3-м сообщении, подводится итог этих исследований. Дифрактометрические, электронографические и электронномикроскопические исследования исходных, промежуточных и конечных фаз, представленных главным образом слоистыми силикатами и алюмосиликатами, свидетельствуют о кристаллохимической преемственности структур при преобразовании каолинита в монтмориллонит, слюды и смешаннослойные фазы (ректорит, тосу-

[1]) Die Arbeit wurde im Röntgenlaboratorium des Lehrstuhls für Kristallographie und im Laboratorium für hohe Drucke und Temperaturen des Instituts für Erdkrustenforschung der Staatlichen Universität Leningrad angefertigt.

дит и пр.). Это проявляется в сохранении мотива структуры в плоскости базиса, в характере заселения октаэдрических позиций, в стремлении к сохранению последовательности чередования и наложения слоев (политипии) и в деформационных процессах, испытываемых структурами исходных минералов при такого рода преобразованиях. Широкое участие различных упорядоченных и неупорядоченных смешаннослойных образований в изучаемых реакциях является следствием трансформационного пути преобразования структур. При этом на ход их формирования и тип образуемой постройки существенное влияние оказывает химизм среды. Таким образом, механизм трансформационных термальных преобразований алюмосиликатов, давно известный для ,,сухих" систем, в значительной мере проявляется в сложных, поликомпонентных системах под давлением H_2O и характеризуется выявленными в работе специфическими индивидуальными чертами.

1. Einführung

Bei der Untersuchung der Strukturumwandlungen des Kaolinits und Metahalloysits unter hydrothermalen Bedingungen in Anwesenheit der Chloride von K, Na, Ca, Mg wurde gefunden (FRANK-KAMENETZKIJ u. a. 1966, 1968, 1969), daß in den Syntheseprodukten bei all ihrer Vielfalt die bekannte Aufeinanderfolge einiger Strukturbesonderheiten der Ausgangsmaterialien zu beobachten ist, was mit der Ähnlichkeit der Kristallstrukturen, die an der Reaktion beteiligt sind, und mit den Besonderheiten der im festen Zustand ablaufenden Prozesse verbunden ist. Im vorliegenden Artikel werden die Grundzüge solcher Strukturfolgen verglichen, die für die Transformationsumwandlungen der Kaolinminerale typisch sind. Zu den Ausgangsmineralen gehören Kaolinit (Prosjanowsker Lagerstätte/UdSSR), Metahalloysit (Michalowezker Lagerstätte/UdSSR), Dickit (Turomginsker Lagerstätte von Alunit/RSFSR).

Die Minerale wurden mit den Chloriden von K, Na, Ca, Mg (p. A.) im Verhältnis 2 : 1 gemischt und in dicht verschlosesnen Platinampullen bei $p_{H_2O} = 900 \pm 25$ kg cm^{-2} und $T = 200-500 \pm 10$ °C im Reaktor mit kaltem Verschluß gehalten (modifizierte Apparatur nach TUTTLE 1948). Die Strukturcharakteristiken der Neubildungen wurden durch Kombination von Diffraktometrie (DRON-I mit Szintillationszähler)[1]), Elektronographie (EM-4) und Elektronenmikroskopie genügend hoher Auflösung unter Benutzung der Mikrodiffraktion (JEM-6A)[2]) ermittelt.

Weiter unten werden die Grundstrukturgesetzmäßigkeiten der durch Transformation umgewandelten Kaolinminerale betrachtet, die bei der Untersuchung des reichlichen experimentellen Materials hervortraten, das im wesentlichen veröffentlicht wurde (GOILO u. a.; FRANK-KAMENETZKIJ u. a. 1969).

2. Strukturmotiv der Basisfläche

Umwandlungen des Kaolinits in fester Phase unter hydrothermalen Bedingungen bei Anwesenheit der Chloride von K, Na, Ca, Mg (Fig. 1) zeigen, daß die Zwischen- und Endprodukte im wesentlichen von Schichtsilikaten und Alumosilikaten mit ähnlichen Basismotiven gebildet werden. Sogar eine Erhöhung der Reaktionszeit auf drei volle Tage und eine Erhöhung der Temperatur von 200

[1]) Aufnahmebedingungen: Cu-K_α, Ni-Filter, 35 kV, 24 mA, 1°/1200, RS = 1 – 2 s.

[2]) Die Aufnahmen wurden angefertigt im Laboratorium für Elektronenmikroskopie des WSEGEI.

auf 500 °C führten nicht zur Bildung nichtschichtiger Phasen (Zeolithe, Feldspäte u. ä.), die sich unter analogen Bedingungen bei der Synthese aus Gläsern und Gelen bilden (BARRER, DENNY). Nur an der Grenze der Temperaturbeständigkeit des Kaolinits, wenn er besonders schnell zerstört wird, zeigen sich Spuren der andalusitähnlichen Phase ($T = 475$ °C, 3–22 h).

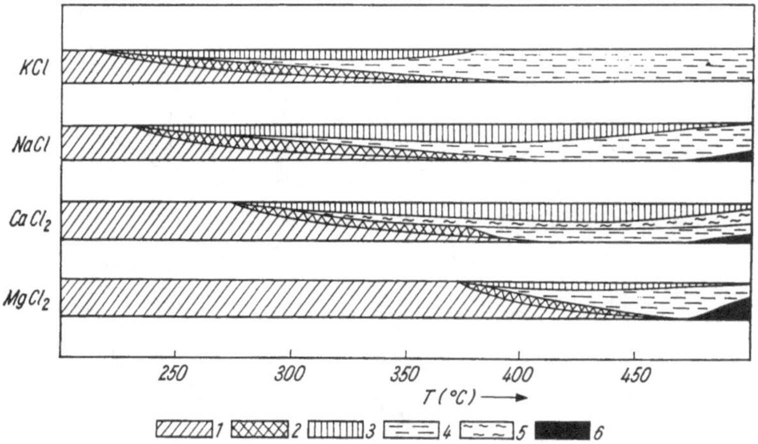

Fig. 1. Assoziation der Phasen im System Kaolinit–Chloride von K, Na, Ca, Mg(p_{H_2O} = 900 kg cm^{-2}, 200 – 500 °C, 22 h)

1 – Kaolinit 1 T; *2* – Kaolinit-“disord”.; *3* – Montmorillonit; *4* – K, Na-Hydroglimmer (aus Versuchen mit KCl und NaCl), Glimmer–Montmorillonit 1:1 (mit $CaCl_2$) und Chlorit–Montmorillonit 1:1 (mit $MgCl_2$); *5* – hexagonales Anorthitanalogon; *6* – andalusitähnliche Phase

Die Spezifik der synthetisierten Phasen zeigt sich in verschiedenen Verbindungen der oktaedrischen und tetraedrischen Schichten mit der Bildung von 2 Schicht-, 3 Schicht-, 4 Schicht- und Vielschichtstrukturen. In den Diffraktogrammen für alle schichtigen Phasen, die an der Reaktion beteiligt sind, zeigt sich das im Auftreten der Reflexe hk0 (h, k = 0, 1, 2, 3, . . ., wobei h, k, h + k gerade Zahlen sind) und in wesentlichen Änderungen der Reflexlage 001 und hkl (z. B. die Versuche mit KCl und $CaCl_2$ – Fig. 2, 3).

Nicht unwesentliche Varianten des Parameters b ($\Delta b = \pm 0{,}06$ Å), die fixiert werden durch Verschiebung des Maximums von 060, zeugen von einer unbedeutenden Änderung des Strukturmotivs der Basisfläche.

Es ist interessant, die Reihenfolge der Ausscheidung der Phasen zu betrachten, die bei Kaolinitmineralen unter hydrothermalen Bedingungen unter Zusatz von $CaCl_2$ entstehen und die aus Gläsern und Gelen analoger Zusammensetzung synthetisiert wurden. Im ersten Falle entstehen aus Kaolinit zwei grundlegende Phasen, eine quellfähige – 14 Å und eine nicht quellfähige – 7 Å (Fig. 3). Die quellfähige Phase, die röntgenographisch nach verschiedener Behandlung untersucht wurde, erwies sich als Montmorillonit (möglicherweise mit tetraedrischem Al). Im Laufe der Zeit und bei Erhöhung der Synthesetemperatur von 400 auf 500 °C wandelt sich diese Phase um in eine geordnete gemischtschichtige Phase von Typ Glimmer-Montmorillonit (Fig. 3). Die 7 Å-Phase wurde elektronographisch (Tab. 1) bestimmt als einschichtige hexagonale Phase analog dem

Anorthit; sie erscheint bei 300–500 °C nach 22 h (Fig. 3); nach einer Versuchsdauer von 70 h bei 500 °C verändert sie sich in eine zweischichtige Anorthitphase.

Die Synthese der zweischichtigen Phase mit Anorthitzusammensetzung wurde unter analogen Bedingungen erreicht (BARRER, DENNY) aus Gläsern und Gelen

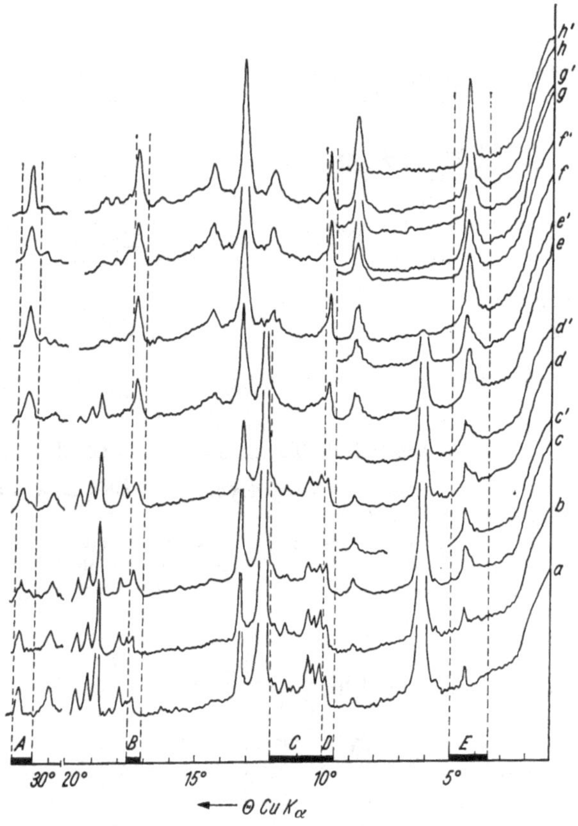

Fig. 2. Aufeinanderfolge der Strukturmotive bei Umwandlung von Kaolinit in Glimmer bei Anwesenheit von KCl (p_{H_2O} 900 = kg cm^{-2}, 200–500 °C, 22–44 h).
Diffraktionskurven: *a* – 200 °C, 22 h; *b* – 250 °C, 22 h; *c* – 300 °C, 22 h; *d* – 300 °C, 44 h; *e* – 350 °C, 22 h; *f* – 400 °C, 22 h; *g* – 450 °C, 22 h; *h* – 500 °C, 22 h; (*c'*–*h'* – die Proben *c*–*h* mit Äthylenglykol gesättigt).
Gebiete der Peaks der schichtigen Phasen: *A* – 060; *B* – 200, 130; *C* – Gebiet, das den Übergang Koalinit 1 T → "Disord." charakterisiert; *D* – 020; *E* – 001 des Hydroglimmers

bei Temperaturen < 350 °C, oberhalb dieser Temperatur kristallisierte Anorthit, und die quellfähigen Phasen zeigten sich nicht. So bilden sich noch bei 400 bis 500 °C metastabile schichtige Phasen mit Anorthitzusammensetzung durch Transformationsumwandlungen aus Kaolinit, während sich aus Gläsern und Gelen ähnlicher Zusammensetzung unter analogen Bedingungen nur der stabile Anorthit bildet (GOLDSMITH, EHLERS). Ungeachtet der Ähnlichkeit der chemischen Zusammensetzung der 7 Å-Phase und des Anorthits sind ihre Strukturen verschieden: die erste ist ein Schichtsilikat (TAKEUCHI, DONNAY) und

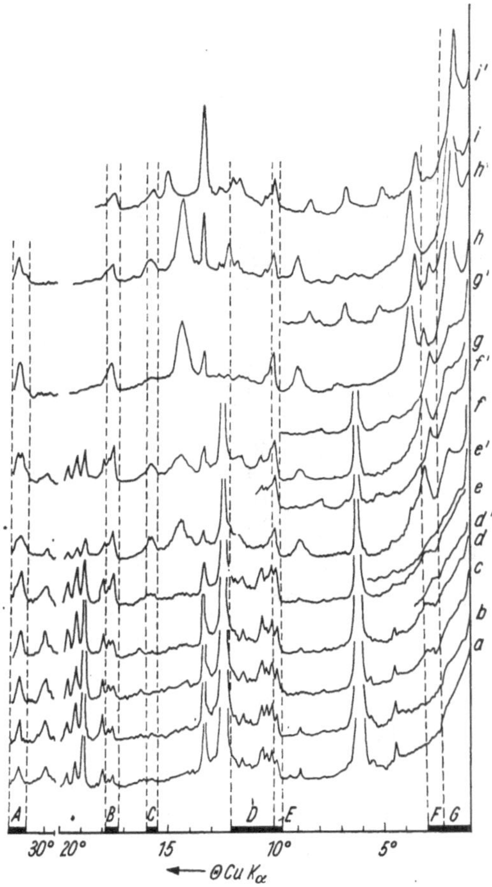

Fig. 3. Aufeinanderfolge der Strukturmotive bei der Umwandlung des Kaolinits in Rektorit und anorthitähnliche Phase in Gegenwart von $CaCl_2$ (p_{H_2O} = 900 kg cm^{-2}, 200 – 500 °C, 22 – 44 h)

Diffraktionskurven: *a* – 200 °, 22 h; *b* – 250°, 22 h; *c* – 300°, 22 h; *d* – 300°, 44 h; *e* – 350°, 22 h; *f* – 400°, 44 h; *g* – 400°, 22 h; *h* – 450°, 22 h; (*i* – 500°, 22 h (*d'* – *i'* – die Proben *d* – *i* mit Äthylenglykol gesättigt).

Gebiete der Peaks der schichtigen Phasen: *A* = 060; *B* – 200, 130; *C* – 022 hexagonales Anorthitanalogon; *D* – Gebiet, das charakteristisch ist für den Übergang 1 T → "Disord." des Kaolinits; *E* – 020; *F* – 001 des Montmorillonits; *G* – 001 der Phase Glimmer–Montmorillonit 1:1

wahrscheinlich in der Struktur analog dem Kimrit (Kaschajev), und die zweite ein typisches Gerüstalumosilikat. So ist das Erscheinen der quellfähigen Phasen und des schichtigen Analogons des Anorthits im gegebenen Falle verbunden mit der Spezifik der Transformationsumwandlung von Kaolinit unter Beibehaltung der Strukturgegebenheiten. Die quantitativen Wechselbeziehungen der quellfähigen und nicht quellfähigen Phasen (Fig. 3, 4) zeigen die Tendenz zu verschiedener Bildung in Abhängigkeit von der Versuchsführung und dem strukturellen Aufbau des Ausgangsmaterials. So wird bei Erhöhung der Temperatur (Fig.3, bei 450–500 °C) eine Zunahme des Gehaltes an 7 Å-Phase bei

Tabelle 1

Elektronographische Daten für die anorthitähnliche Phase, synthetisiert im System Kaolinit+$CaCl_2$
(p_{H_2O} = 900 kg cm^{-2}, 350–500 °C, 22 Stunden)

I	d(Å)	hkl	I	d(Å)	hkl
80	4,43	020	60	1,52	242
80	3,78	021	70	1,48	204
100	2,82	022	100	1,47	060
100	2,55	200	10	1,41	044
65	2,41	201	10	1,38	243
35	2,20	040	20	1,36	062
35	2,16	023	55	1,27	400
50	2,11	041	2	1,23	244
25	2,09	202	10	1,23	401
5	1,89	042	30	1,22	205
50	1,76	203	50	1,22	420
40	1,67	240	15	1,20	421
5	1,64	043	15	1,16	422
60	1,62	241	35	1,15	064

a = 5,10 Å; b = 8,84 Å; c = 7,28 Å; $\alpha = \beta = \gamma = 90°$

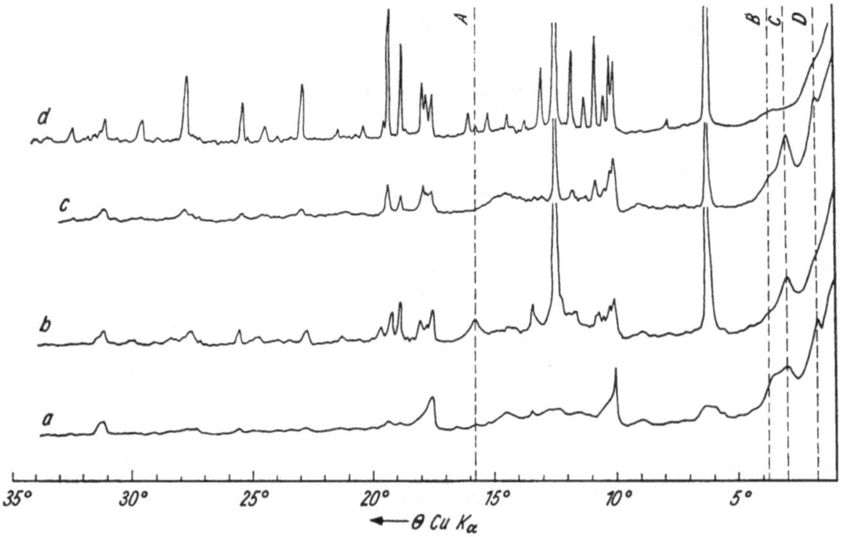

Fig. 4. Einfluß des Kristallinitätsgrades des Ausgangsmaterials auf die Bildung des Rektorits und der anorthitähnlichen Phase in den Systemen Kaolinit, Dickit + $CaCl_2$ (p_{H_2O} = 900 kg cm^{-2}, 400 °C, 22 h).
Mechanisch zerkleinert (10 min): *a* – Kaolinit; *c* – Dickit.
Nicht zerkleinert: *b* – Kaolinit; *d* – Dickit.
Bezeichnung der Peaks: *A* – hexagonales Anorthitanalogen; *C* – 001 des Montmorillonits; *B* – 002, *D* – 001 der Phase 1:1 Glimmer–Montmorillonit

Verringerung der quellfähigen Phase beobachtet. Bei zerriebenem Kaolinit (Vergrößerung des Anteils an spaltbarer Oberfläche) bilden sich nur quellfähige Phasen (Montmorillonit und Glimmer: Montmorillonit 1:1), während bei nicht zerriebenem Kaolinit sich sowohl die quellfähige als auch die nicht quellfähige 7 Å-Phase bildet (Fig. 6a, b). Analoge Resultate werden erhalten bei der Bearbeitung von Dickit (Fig. 4b, 4c), aber in diesem Falle verläuft die Reaktion weniger intensiv, was zusammenhängt mit den unterschiedlichen Korngrößen des Kaolinits und des Dickits. An Hand der Ergebnisse der Elektronenmikroskopie mit Mikrodiffraktion (FRANK-KAMENETZKIJ u. a. 1969) wird deutlich gezeigt, daß die quellfähigen Phasen sich von den äußeren Teilen der Kaolinitkristalle her entwickeln. Das Erscheinen der 7 Å-Phase als hexagonales Skelettgebilde erlaubt zu sagen, daß sie sich am schnellsten an der Oberfläche der Kaolinitkristalle bildet und die Bildung dieser zwei verschiedenen Phasen kontrolliert wird durch das Wechselverhältnis der Korngrenzen- und Spaltoberflächen der Kristalle.

Die röntgenographischen und elektronenmikroskopischen Daten der Kaolinitumwandlung unter hydrothermalen Bedingungen in Gegenwart von $CaCl_2$ zeigen mit Bestimmtheit die Aufrechterhaltung der Ausgangsstruktur in den Neubildungen der qellfähigen Phasen, die für die Synthese aus Kolloiden und Gläsern unter analogen Bedingungen nicht typisch sind; ebenso in den Bildungen der hexagonalen Phase der Anorthitzusammensetzung unter Bedingungen, unter denen sich aus Gläsern Anorthit bildet. Alles Gesagte ebenso wie die Ähnlichkeit der Strukturmotive und die kurzen Versuchszeiten geben die Mölichkeit anzunehmen, daß die Synthese dieser Phasen im festen Zustand bei Strukturumwandlung des kristallinen Materials unter dem Einfluß mineralisierender Lösungen vonstatten geht. Es ist möglich, folgende Wege der Strukturumwandlung der Kaolinitmineralien unter hydrothermalen Bedingungen in Anwesenheit von $CaCl_2$ zu postulieren (siehe Fig. 7):

1. Kaolinit → 7 Å-hexagonales Analogon des Anorthit → Anorthit. Diese Phasen bilden sich von der Oberfläche der Kristalle her unter Beibehaltung des Kieselsäuretetraedernetzes $(Si, Al)_2O_5$.
2. Kaolinit → quellfähige Phasen (Montmorillonit → 1:1 geregelte gemischtschichtige Phase Glimmer-Montmorillonit). Die Phasen bilden sich von den Kristallgrenzen her, dabei wird das Strukturmotiv der Basis $(Si, Al)_2O_5$ ebenso wie bei der 7 Å-Phase erhalten, aber auch das dreidimensionale Gerüst ist möglich.

3. Das Bestreben zur Erhaltung der Aufeinanderfolge der Schichten (das Weiterbestehen der Polytypie)

Die vollkommene Regelung der Pakete in der Basisebene für die oben beschriebenen quellfähigen Phasen, die sich von den Korngrenzen der Kaolinitkristalle her bilden, gibt nicht die Möglichkeit, den Charakter der Veränderung der Gitterstruktur zu ermitteln. Für das Auffinden der Gesetzmäßigkeiten eines solchen Typs wurden für die Umwandlung der Kaolinitmineralien in KCl-Lösung spezielle Versuche mit kürzeren Versuchszeiten (3 h) durchgeführt. Dabei erscheint, wie zu erwarten war, dioktaedrischer Glimmer, der sich über die dazwischenliegende ungeregelte gemischtschichtige Phase von Glimmer-Montmorrillonit bildet (Fig. 2). Die röntgenographische und elektronographische Analyse der

Tabelle 2

Elektronographische Daten (02l, 11l) für Glimmer, synthetisiert aus Dickit (Probe 1), Kaolinit (Probe 2) und Metahalloysit (Probe 3)

Probe 1 $2 M_I > 1 M$			Probe 2 $1 M < 2 M_1$			Probe 3 „Disord."		
hkl ($2 M_1$)	d(Å)	I	hkl (1 M)	d(Å)	I	hkl	d(Å)	I
020	4.47	100	020	4.48	100	020	4.47	100
—	—	—	$\bar{1}11$	4.34	50	—	—	—
022	4,085	35	021	4.09	35	022	4.08	10
$\bar{1}13$	3,86	45		3.86	10	$\bar{1}13$	3.88	15d
113	3,66	55	$\bar{1}12$	3.65	60	113	3.67	35d
$\bar{1}14$	3.49	40		3.48	8	$\bar{1}14$	3.50	5
024	3.34	50	022	3.34	50	024	3.35	5
114	3,19	40		3,21	8	114	3.20	5
$\bar{1}15$	3.06	65	112	3,06	65	$\bar{1}15$	3.04	40d
025	2,99	15	$\bar{1}13$	2.93	40	—	—	—
a = 5,18 Å			a = 5,18 Å			a = 5,18 Å		
b = 8,94 Å			b = 8,95 Å			b = 8,94 Å		
c = 19,85 Å			c = 10,3 Å					
β = 95° 40′			β = 101° 20′					

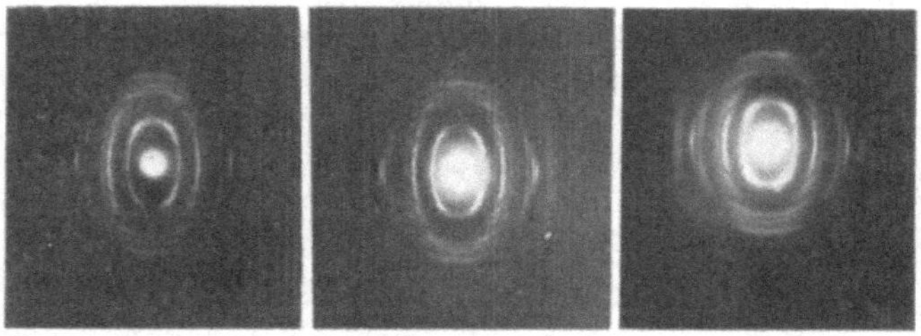

Fig. 5. Elektronogramme polymorpher Glimmer mit schräger Textur, aufgenommen mit EM-4 (Kohle 45° φ)

synthetisierten Glimmer zeigte, daß ihre polytypen Modifikationen in dem bekannten Ausmaß von der Ausgangsstruktur kontrolliert werden (Tab. 2, Fig. 5, 6). So bildet sich über Metahalloysit geregelter Hydroglimmer vom Illittyp, und über Kaolinit (1T) bildet sich Glimmer mit deutlich überwiegendem Anteil der Modifikation 1M über 2 M_1, über Dickit ($2M_1$) bildet sich Glimmer mit überwiegender Modifikation $2M_1$ über 1M.

In den zu untersuchenden Umwandlungen der Kaolinit- in Hydroglimmerstrukturen ist zu beachten, daß die Struktur im Ganzen wesentliche Veränderungen durchmacht (Zweischichtsilikat → Dreischichtalumosilikat), jedoch die Zentren der Ditrygontetraedergitter benachbarter Schichten behalten das grundlegende Motiv der Überlagerung bei, was die Kontinuität des dreidimensionalen Motivs der Ausgangsstruktur in den Neubildungen beweist.

3.1 Oktaedrische Schicht

Die oben betrachteten Beispiele behandeln die Kontinuität der strukturellen Züge bei Umwandlungen von Kristallen ohne Veränderung des Charakters der Anordnung der oktaedrischen Positionen. Wenn bei Zugabe von K, Na und Ca die Beibehaltung des dioktaedrischen Motivs bei Umwandlungen von Kaolinitmineralen sich nicht als ungewöhnlich erweist, dann fordert das Erscheinen von dioktaedrischen Motiven bei Zusatz von Mg besondere Aufmerksamkeit. Der kleine Ionenradius des Mg und die weite Verbreitung der Mg-reichen trioktaedrischen Glimmer, Chlorite, Montmorillonite u. a. geben die Grundlage, das Auftreten derartiger Bildungstypen von bei Zutritt von $MgCl_2$

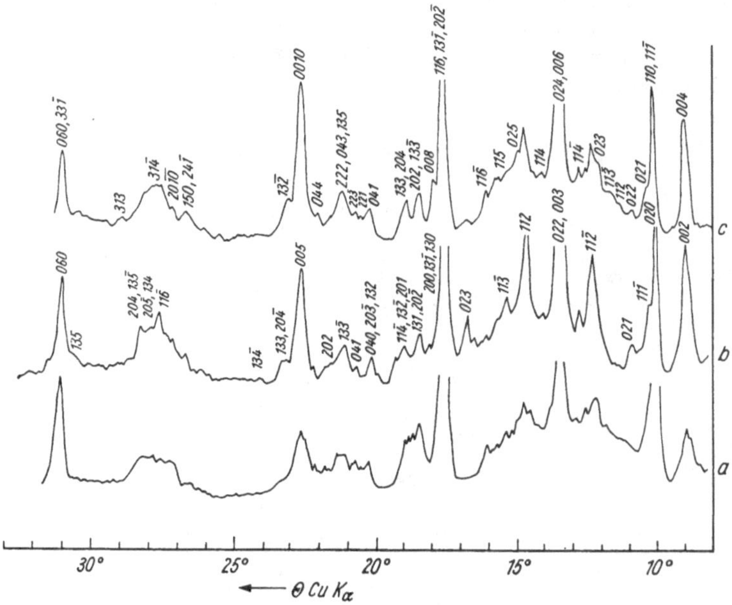

Fig. 6. Das Bestreben der Beibehaltung der Aufeinanderfolge der Wechselschichten („Vererbung" der Polytypie) bei der Umwandlung der Minerale der Kaolinitgruppe in Anwesenheit von KCl (p_{H_2O} = 900 kg cm^{-2}, 500 °C, 3 h).

Diffraktogramme der Polytypie der Glimmer, die gebildet wurden aus: *a* – Metahalloysit ("disordered"); *b* – Kaolinit (überwiegend 1 M, Beimischung von 2 M_1); *c* – Dickit (überwiegend 2 M_1, beigemischt 1 M)

zum System zu erwarten. Faktisch erscheint in diesem Falle die pseudomorphe Umwandlung des Kaolinits zu dioktaedrischem Montmorillonit. Mit Erhöhung der Versuchsdauer (von 3 bis 70 h) und der Temperatur (von 400 bis 500 °C) geht dieser Montmorillonit in eine geregelte gemischtschichtige Phase Chlorit–Montmorillonit (1:1) über (Frank-Kamenetzkij u. a. 1969). Der Charakter der Schichtanordnung der Kationen in dieser Phase konnte bei gleichmäßiger Synthese über 16 Ordnungen von 001 festgestellt werden für eine Probe, die mit Glyzerin gesättigt war und bei 550 °C erhitzt wurde (Fig. 7).

Das Verhältnis der Peakhöhen der Elektronendichten gibt Auskunft über die dioktaedrische Natur dieser gemischtschichtigen Struktur, wo Mg im wesentlichen in die Brucitschicht und in die Zwischenschichträume eingeht; im dreischichtigen Paket wird die Dioktaedrigkeit im wesentlichen aufrechterhalten. Diese Schlußfolgerung wird bestärkt durch eine unbedeutende Kontraktion um 1 Å in c^* Richtung der geregelten gemischtschichtigen Phase bei ihrer Behandlung mit LiCl mit anschließender Erhitzung und Sättigung mit Glyzerin nach Green-Kelley.

So wird im gegebenen Falle der geregelte dioktaedrische Chlorit-Montmorillonit synthetisiert, ähnlich den früher beschriebenen (Sudo, Kodama; Sudo) natürlichen gemischtschichtigen Bildungen in Japan, deren Analoga aus Aluschtiten der Krim detailliert kristallchemisch untersucht worden sind (Frank-Kamenetzkij u. a. 1965; Frank-Kamenetzkij u. a. 1968) und die Tosudit genannt wurden.

Es ist möglich, daß sich im Verlauf der Zeit unter günstigen Bedingungen die dioktaedrische Schicht mit einer solchen Struktur in eine trioktaedrische durch die Migration von Mg aus quellfähigen Montmorillonitzwischenschichten umwandeln kann.

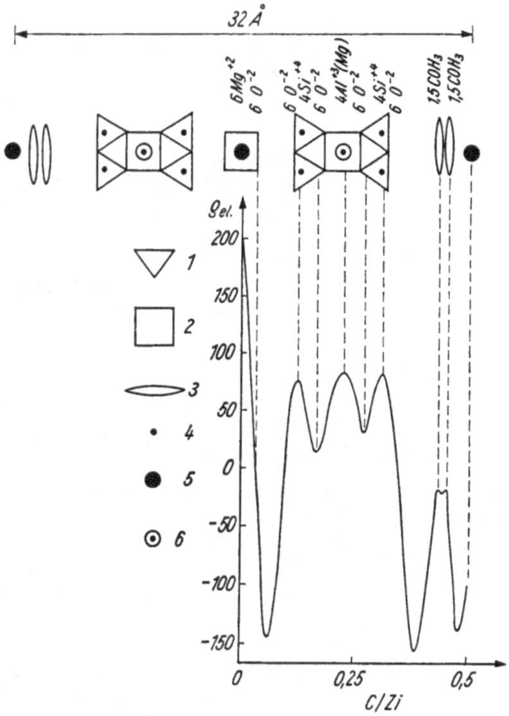

Fig. 7. Strukturschema der dioktaedrischen, geordneten, synthetischen Variation des Tosudits, die aus Kaolinit in Anwesenheit von $MgCl_2$ (p_{H_2O} = 900 kg cm^{-2}, 450 °C, 22 h) erhalten wurde

Bedeutung: *1* – Tetraeder; *2* – Oktaeder; *3* – Glyzerinmoleküle; *4* – Si; *5* – Mg; 6 – Al

3.2 Die Strukturveränderung des Kaolinits durch Umwandlungsprozesse

Bei der Untersuchung der Phasentransformationsmetamorphose des Kaolinits wurde festgestellt, daß die Struktur des Kaolinits, der noch nicht in Reaktion getreten war oder noch nicht völlig durchreagiert war, nicht unverändert bleibt und gesetzmäßige Veränderungen durchmacht. Bis zum Beginn der Bildung einer neuen Phase (bis ~ 250 °C unter Zugabe von KCl oder bis ~ 400°C unter Zusatz von $MgCl_2$) kristallisiert der Kaolinit unter Strukturverbesserung um. Aus den Diffraktogrammen und den Elektronenaufnahmen zeigte sich dies in einer besseren Entwicklung der Reflexgruppen $020-1\bar{1}\bar{1}$, $02\overset{\pm}{\bar{1}}$, $1\bar{3}0-131$ ... und in einer Vergrößerung der Zahl der Kristallite mit regelmäßigen hexagonalen Konturen. Beim Erreichen der Anfangstemperatur der Bildung irgend einer neuen Phase in Gegenwart einer der eingesetzten Kationen wurde der Übergang des geregelten triklinen Kaolinits in einen pseudomonoklinen Kaolinit bemerkt, der geordnete Basisschichten aufweist (Fig. 1).

Die geordneten Kaolinitstrukturen und die Verkleinerung der Kristallabmessungen bei Zugabe von KCl und $CaCl_2$ (in 22 h) bildeten sich schon bei 250—300 °C (Fig. 2, 3), d. h. bei einer bedeutend niedrigeren Temperatur als unter den „reinen" Wasserdampfdruckbedingungen ($\sim$ 400 °C) oder in Anwesenheit von $MgCl_2$ (auch bei $\sim$ 400 °C). Es ist danach zu bemerken, daß eine Verschlechterung des Kristallinitätsgrades und der Regelung der Kaolinitstruktur unter hydrothermalen Bedingungen in Anwesenheit der Chloride von K, Na, Ca, Mg die Bildung neuer Phasen (Montmorillonit, ungeordnete und geordnete gemischtschichtige Phasen) nach sich zieht. Analoge Strukturveränderungen des Kaolinits, nämlich Ordnung der Basisebene, gehen vonstatten (Goilo u. a.) unter dem Einfluß gerichteten Druckes (stress); ein der Größe nach ähnlicher hydrostatischer Druck bewirkt keinen solchen Typ der Störungen (bis 400 °C bei $p_{H_2O} = 900$ kg cm^{-2} in 22 h). So führen der gerichtete Druck und auch die Kationenzusätze bei Transformationsumwandlungen des

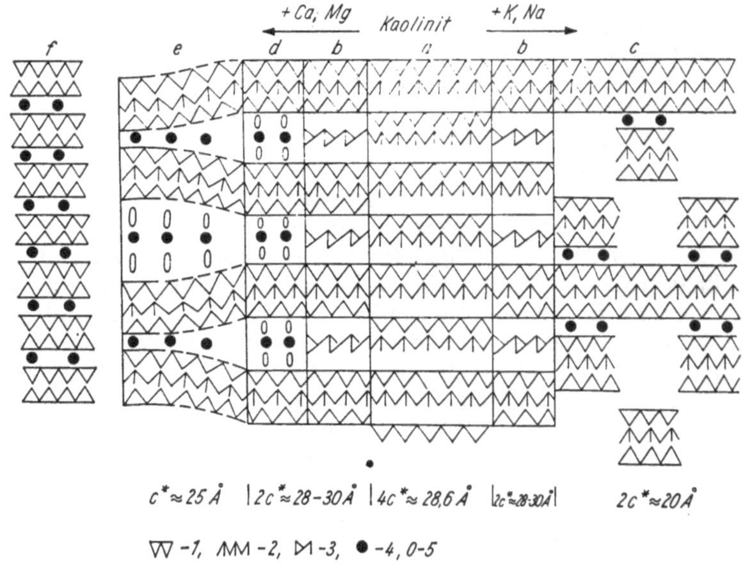

Fig. 8. Schema der Transformationsumwandlung des Kaolinits unter hydrothermalen Bedingungen in Anwesenheit der Chloride von K, Na, Ca, Mg ($p_{H_2O} = 900$ kg cm^{-2}, 200—500 °C, 22—70 h) in Glimmer, gemischtschichtige Phasen und einschichtiges hexagonales Anorthitanalogon.

Schematische Darstellung der Strukturmotive: *a* — des Kaolinits; *b* — der Beidellitähnlichen Phase; *c* — des Hydroglimmers (K, Na); *d* — des Mg, Ca-Montmorillonits; *e* — des Glimmer-Montmorillonits 1:1 (Ca) oder des Chlorit–Montmorillonits 1:1 (Mg); *f* — des hexagonalen Anorthitanalogens.

Bezeichnungen: *1* — Tetraeder, *2* — Oktaeder, *3* — teilweise zerstörte Oktaeder, *4* — Zwischenschicht-Kationen, *5* — Wassermoleküle

Kaolinits zu einem ähnlichen Effekt, nämlich der Ordnung seiner Struktur. Der Beginn der Ordnung und die Anfangstemperatur der Synthese der neuen Phasen werden bestimmt durch den Charakter der eingeführten Kationen (Fig. 1). Der Prozeß der Ordnung der Schichten, der hier betrachtet wird, hat am ehesten Versetzungsnatur; der Mechanismus solcher Verschiebungen durch Aufspaltung der Randversetzungen in Teilversetzungen wurde bei Strukturumwandlungen des Lisardits durch DRIZ beobachtet. Im Falle des Kaolinits wird die Deformation (Kontraktion oder Dehnung senkrecht zur Basis) stimuliert durch die Einführung von Kationen. Der schematisch unterschiedliche Charakter der Umwandlung der Kaolinitstruktur über einen dazwischengeschalteten Montmorillonit-ähnlichen Bau bei Zugabe von K, Na (rechter Teil) und Ca, Mg (linker Teil) ist dargestellt in Figur 8.

4. Bildungsbedingungen gemischtschichtiger Phasen

Das weitverbreitete Auftreten gemischtschichtiger Phasen, die gewöhnlich als Zwischenbildungen bei der Strukturumwandlung des Kaolinits auftreten, folgt unmittelbar aus dem Transformationscharakter des Prozesses und aus dem einheitlich schichtigen Strukturtyp. Wie früher gezeigt wurde, ist dieser Prozeß wesentlich mit der chemischen Natur des umgebenden Mediums verbunden. Die Reaktionen unter Beteiligung von Kaolinit und den Chloriden von K, Na, deren Endergebnis die Synthese der Glimmer mit Muskovit- oder Paragonitzusammensetzung ist, durchlaufen Etappen der Entstehung zwischengeschalteter nichtgeordneter gemischtschichtiger Bildungen vom Typ Glimmer–Montmorillonit. In Anwesenheit von Zusätzen der Chloride von Ca und Mg zum Kaolinit bilden sich über eine dazwischenliegende Montmorillonitphase geregelte gemischtschichtige Phasen 1:1 vom Typ Glimmer–Montmorillonit (Ca) und Chlorit–Montmorillonit (Mg). Man kann behaupten, daß sich bei günstigen Bedingungen diese gemischtschichtigen Phasen in Vermiculit, Chlorit, Clintonit, Margarit u. a. umbilden werden unter Beibehaltung der grundlegenden Elemente des Motivs der Ausgangsstruktur. Der Bildungsmechanismus der schichtigen Phasen aus Kaolinit in unseren Versuchen wird unter Berücksichtigung der Voraussetzung der Kaolinitumwandlung unter Wasserdampfdruck ohne Zusätze betrachtet. Wie ROY und OSBORN zeigten, ist die erste Phase, die sich aus Kaolinit bildet, Beidellit-ähnlich, die im Laufe der Zeit in Pyrophyllit übergeht (ALTHAUS). Wenn man die bereits oben betrachteten kristallchemischen Erwägungen über die Transformation der Strukturen (Fig. 8) verwendet, kann man die Bildungsreaktion der Beidellit-ähnlichen Phase hier wie folgt schematisch ausdrücken:

$$2\,[Al_2Si_2O_5(OH)_4] \xrightarrow{T,\,p_{H_2O}}$$

$$[Al_{2-x}(OH)]_{6-3x}]\,Al_2 \underset{n R^+ aq}{\downarrow} [Si_{4-x}Al_xO_{10}]\,(OH)_2 + x\,SiO_2 + n\,a\,q^{1)}$$

4.1 Zusätze von K und Na

Bei diesen Zusätzen muß die sich bildende dreischichtige 14 Å-Beidellit-ähnliche Struktur als Zwischenprodukt betrachtet werden; die hohe Bildungsgeschwindigkeit von Glimmer aus Kaolinit bei verhältnismäßig niedriger Temperatur ist

[1]) Für die Beidellit-ähnliche Phase ist $x < 2$.

in diesem Falle leicht zu erklären durch das starke Schrumpfen der Beidellit-ähnlichen Struktur in c^* Richtung (~ 30%), was begleitet wird von einem intensiven Kornzerfall mit Erneuerung der reagierenden Oberflächen. Die Pyrophyllit-ähnliche Phase zerfällt dabei in Blöcke, die durch Hohlräume getrennt sind; je Block 3 (oder 2) Glimmerpakete (Fig. 8). Die gebildeten Hohlräume füllen sich mit Material, das herausgetragen wird aus teilweise zerstörten oktaedrischen und tetraedrischen Schichten, wodurch sich eine defekte Glimmerstruktur bildet, deren Diffraktionsbild der nichtgeordneten gemischtschichtigen Phase vom Typ Glimmer–Montmorillonit entspricht. Na begünstigt im Vergleich mit K in geringerem Maße die Kontraktion der Beidellit-ähnlichen Struktur. Das ist verständlich, da die Größe (r) und die Hydratationsenergie (ΔE) dieser Ionen verschieden sind (für Na: $r = 0{,}98$ Å; $\Delta E = 0{,}25$ kcal/Grammion; für K: $r = 1{,}33$ Å, $\Delta E = -0{,}25$ kcal/Grammion, nach SAMOILOW 1957, S. 161). Deshalb begünstigt die Zugabe von Na die Phasenbildung mit hohem Gehalt an quellfähiger Komponente und führt zu einer geringen Erhöhung der Anfangstemperatur der Glimmersynthese.

Wenn man die Strukturbesonderheiten des Transformationsmechanismus des Kaolinitbaus in einen Glimmerbau berücksichtigt, kann man die Bildungsreaktion z. B. des K-Glimmers über die Beidellitähnliche Phase wie folgt schreiben:

$$3\left\{[Al_{2-x}(OH)_{6-3x}]\,Al_2 \underset{nR^+\cdot aq}{\downarrow} [Si_{4-x}Al_xO_{10}](OH)_2\right\} + 4\,KCl \xrightarrow{T,\,P_{H_2O}}$$

$$3\,\{KAl_2\,[AlSi_3O_{10}](OH)_2 + \cdots + \{KAl_2\,[AlSi_3O_{10}](OH)_2\}$$

$$+\,4\,HCl + 6\,H_2O$$

Der erste Teil der Reaktion ist ein typischer Transformationsteil; er geht vor sich unter Beibehaltung des Motivs der Ausgangsstruktur; der zweite läuft ab in der Dampfphase mit der Kristallisation von Neubildungen oder zwischen den Blöcken oder außerhalb von ihnen. Weil K, Na in die Struktur von den Rändern beliebiger Kristallteilchen eintreten können, ist es schwer, eine bedeutende Regelung der Neubildungen in c^*-Richtung zu erwarten. In Figur 8 sind die Fragmente der wahrscheinlichen Wechsellagerung dargestellt, aus denen sich im weiteren nichtgeordnete gemischtschichtige Bildungen aufbauen. Die Vielfalt solcher Fragmente für die betrachteten Transformationsreaktionen wird kontrolliert durch die Abmessungen in c^*-Richtung der Ausgangs-, der Zwischen- und der Endphasen, durch die Migration der K-Ionen im Zwischenschichtraum. Die Folge davon ist die wahrscheinliche Wechsellagerung von 2—3 Strukturfragmenten. Die Tendenz des Überganges der gemischtschichtigen Bildungen in den geordneten Zustand stimmt mit grundlegenden thermodynamischen Erwägungen überein (SATO u. a.).

4.2 Zusätze von Ca und Mg

Diese führen zu Pseudomorphosen von Montmorillonit nach Kaolinit, die der Synthese von geordneten gemischtschichtigen Phasen vorausgehen (Fig. 3). (Frank-Kamenetzkij u. a. 1969). Im Unterschied zu früheren Feststellungen für die Synthese mit Alkalien ist die Anfangstemperatur der Bildung gemischtschichtiger Phasen hier höher als die Temperatur der Montmorillonitsynthese. Tatsächlich kann man sich in der folgerichtigen Reihe von Transformationsstrukturen einen zwischengeschalteten Montmorillonit vorstellen: Kaolinit → Beidellitähnliche Phase → Montmorillonit.

Die Rolle der Kationenzusätze führt hier, wie man sieht, nur zur Stabilisierung der Beidellit-ähnlichen Phase. Dabei füllen sich ihre Zwischenschichtplätze mit hydratisierten Ca- und Mg-Ionen, was nicht zu einer bedeutenden Kontraktion der Struktur in c^*-Richtung führt (Fig. 8). Das passive Verhalten der Ca- und Mg-Ionen im Vergleich zur aktiven Rolle des K und Na bei verhältnismäßig niedrigen p, T-Parametern der Versuche erklärt sich am ehesten durch die große Bedeutung ihrer Hydratationsenergien (E für Ca $= +0{,}45$ kcal/Grammion; für Mg $+2{,}61$ kcal/Grammion, Samoilov 1957, S. 161). Nur bei einer Temperatur von $\sim$ 400 °C (22 h) wird die folgerichtige Reihe Kaolinit → Beidellit-ähnliche Phase → Montmorillonit bei Anwesenheit von $CaCl_2$ und $MgCl_2$ abgeschlossen durch die Bildung geordneter gemischtschichtiger Phasen. Schematisch kann man z. B. für $CaCl_2$ folgende Transformationsreaktion formulieren:

$$4\,\{[Al(OH)_3]Al_2\, n\, R^+ \cdot a\, q\, [Si_3AlO_{10}](OH)_2\} + 3\,CaCl_2 \xrightarrow{T,\, p_{H_2O}}$$

$$2\,\{[Ca_{0,5} \cdot n\, H_2O]Al_2\, [Si_3Al\,O_{10}](OH)_2 \cdot (?)\, CaAl_2\, [Si_2Al_2O_{10}]\,(OH)_2\} + \ldots$$

$$+ Al_2SiO_5 + 6\,HCl + SiO_2 + n\, a\, q\,.$$

Der erste Teil der Reaktion ist ein typischer Transformationsteil. Der zweite betrifft Al und Si, die im Verlaufe einer solchen Umwandlung teilweise aus der Struktur herausgelöst werden unter Bildung beigemischter hochtonerdehaltiger Phasen: Böhmit, Diaspos, bei höheren Temperaturen die metastabile andalusitähnliche Phase. Die Gegenüberstellung der Gitterparameter (c^*) für die Phasen, die an den Transformationsreaktionen teilnehmen, zeigt, daß die Bildung von geordneten gemischtschichtigen Phasen in diesen Fällen bei Zusatz von Ca und Mg zu einer geringeren Deformation der Ausgangsstruktur führt als im Falle der K- und Na-Zugabe. Dieses letztere führt nicht zu einer Verkleinerung der Kaolinitpakete und stimuliert nur eine geringe Verzerrung ihrer Struktur (Übergang 1 T → „Disord.").

Im Unterschied zu Ca führt die Zugabe von Mg zu einer Umwandlung der dazwischenliegenden Phase in einen geregelten gemischtschichtigen Chlorit-Montmorillonit. In diesem Falle sind die Unterschiede der Parameter c^* für die Phasen, die an der Reaktion beteiligt sind, minimal. Die kristallchemischen Besonderheiten von Mg im Vergleich zu Ca können den Unterschied der hier entstehenden gemischtschichtigen Strukturtypen erklären. Während Ca mit wachsender Temperatur und Expositionszeit bestrebt ist, in eine Struktur der Beidellit-ähnlichen Phase, aus dem beweglichen Zustand in einen in der Zwischenschicht fixierten überzugehen unter Bildung der Phase Glimmer-Montmorrillonit (1:1), ist Mg unter den gleichen Bedingungen betrebt, eine Brucitschicht zu bilden, wodurch eine geordnete gemischtschichtige Phase Chlorit–Montmorillonit entsteht.

$$2\left\{[Al(OH)_3\; Al_2 \underset{n\, R^+ \cdot a\, q}{\downarrow} [Si_3AlO_{10}](OH)_2\right\}$$

$$+ 5\,MgCl_2 + 4{,}5\,SiO_2 + m\, a\, q \xrightarrow{T,\, p_{H_2O}}$$

$$\{(Mg_{1,5}Al)\; Al_2\, [Si_3AlO_{10}]\, (OH)_8 \cdot (Mg_{0,5} \cdot n\, a\, q)\; Al_2\, n\, R^+ \cdot a\, q\, [Si_3AlO_{10}]$$

$$\times\; (OH)_2\} + Mg_3[Si_4O_{10}](OH)_2 + 0{,}5\; Al_2SiO_5 + 10\; HCl$$

So erlauben die angeführten Daten einige charakteristische Züge der Transformationsumwandlungen der Minerale der Kaolinitgruppe unter hydrothermalen Bedingungen aufzustellen. Die hier aufgedeckten Gesetzmäßigkeiten sind wahrscheinlich auch typisch für die Umwandlungen anderer Typen von Schichtsilikaten aus der Gruppe der Tonminerale in natürlichen Prozessen der Epigenese und der beginnenden Metamorphose, was bekräftigt wird durch die Ergebnisse der Untersuchung natürlicher Bildungen (LUCAS, MILLOT, KOSSOVSKAJA u. a.). Für einen solchen Typ der Prozesse ist es charakteristisch, daß die Veränderung bereits vorhandener Minerale (Kaolinit, Montmorillonit, Hydroglimmer u. a.) in fester Phase vor sich geht unter Bedingungen, unter denen grundlegende Strukturelemente nicht zerstört werden, wobei im Verlauf der Reaktion mit mineralisierenden natürlichen Lösungen ein Transformationsprozeß der Umwandlung dafür typischer vielfältiger ungeordneter und geordneter gemischtschichtiger Strukturen und mit weiter oben betrachteten charakteristischen Zügen der kristallchemischen Aufeinanderfolge vonstatten geht.

5. Zusammenfassung

1. Die Experimente zur hydrothermalen Veränderung der Minerale der Kaolinitgruppe in Anwesenheit der Chloride von K, Na, Ca und Mg erlauben den Transformationscharakter der Umwandlungen festzustellen, wenn die neugebildeten Strukturen einzelne Strukturfragmente der Ausgangsmaterialien übernehmen.

2. Die dazwischenliegenden und resultierenden Schichtsilikate und Alumosilikate werden unabhängig vom Chemismus des Mediums durch allgemeine Züge der kristallchemischen Strukturfolge (Motiv der Basisschicht, die Anordnung der tetraedrischen und oktaedrischen Positionen) und durch das Bestreben zur Erhaltung der Kontinuität der Wechsellagerung (Polytypie) charakterisiert.

3. Ein großer Teil der verschiedenen gemischtschichtigen Phasen ist in den untersuchten Reaktionen die Folge eines Transformationsweges von Strukturumwandlungen. Dabei zeigt sich auf ihrem Bildungsweg der Chemismus des Mediums von wesentlichem Einfluß.

Literatur

ALTHAUS, E.: Naturwissenschaften **53**, 105 (1966)

BARRER, R. M., DENNY, P. J.: J. chem. Soc. Nr. 3 Part X, 983 (1961)

DRIZ, V. A.: Die Natur der Zerstörung der Ordnung in Strukturen einiger Schichtsilikate, Röntgenographie mineralischer Rohstoffe Nr. 5, 94, 1966 (russ.)

FRANK-KAMENETZKIJ, V. A., KOTOV, N. V., GOILO, E. A.: Kristall und Technik **1** (3), 465 (1966)

FRANK-KAMENETZKIJ, V. A., KOTOV, N. V., GOILO, E. A.: Kristall und Technik **3** (4), 643 (1968)

FRANK-KAMENETZKIJ, V. A., KOTOW, N. Y., GOILO, E. A.: Hydrothermalsynthese in den Systemen Halloysit, Kaolinit mit den Chloriden von K, Na, Ca, Mg unter Druck in: Probleme der Petrologie und genetischen Mineralogie, Moskau, 143 1969 (russ.)

FRANK-KAMENETZKIJ, V. A., LOGVINENKO, N. V., DRIZ, V. A.: Über Tosudit und Aluschtit, Miner. Sb. Lwowsk. Staatl. Univ. **1**, Nr. 22, 70, 1968 (russ.)
FRANK-KAMENETZKIJ, V. A., LOGVINENKO, N. V., DRIZ, V. A.: Tosudite — a new mineral, forming the mixed-layer phase in alushtite. Int. clay conference 1963, **2**, 181—186, (1965)
GOLDSMITH, J. R., EHLERS, E. G.: J. Geol. **60** (4), 386 (1952)
GOILO, E. A., KOTOV, N. V., FRANK-KAMENETZKIJ, V. A.: Experimentelle Untersuchung des Stress-Einflusses und des hydrostatischen Druckes bei verschiedenen Temperaturen auf die kristalline Struktur des Kaolinits in: Physikalische Untersuchungsmethoden der Minerale der Sedimentgesteine, Moskau, **123**, 1966 (russ.)
GREEN-KELLEY, R.: J. Soil Sci. **4** (2), 233 (1953)
KASCHAJEV, A. A.: Dokl. A.N.SSSR **169**, (1), 201 (1966) (russ.)
KOSSOVSKAJA, A. G.: Typisierung und genetische Bedeutung gemischtschichtiger Tonminerale in: Physikalische Untersuchungsmethoden der Minerale der Sedimentgesteine, Moskau **163**, 1966
LUCAS, J.: La transformation des mineraux argileuse dans la sedimentation. Etudes sur les argiles du Trias, Mem. Ser. Carte Geol. d'Alsace et Lorraine, (**23**), 1962
MILLOT, G.: Geologie des argiles, Masson, Paris, 1964
ROY, R., OSBORN, E. F.: Amer. Mineralogist **39**, (11—12), 853 (1954)
SAMOILOV, O. J.: Die Struktur wäßriger Lösungen von Elektrolyten und die Hydratation der Ionen, Moskau, 1957 (russ.)
SATO, MITSUO: Nature **208**, N 5005, 70 (1965)
SUDO, T.: Mineralogical study on clays of Japan, Maruzen Co., Ltd. Tokyo 1959
SUDO, T., KODAMA, H.: Z. Kristallogr., **109**, 379 (1957)
TAKEUCHI, Y., DONNAY, G.: Acta Crystallogr. **12**, 465—470 (1959)

(Eingegangen am 26. April 1969, Übers. 1. Oktober 1969)

Anschrift der Verfasser:

Prof. Dr. V. A. FRANK-KAMENETZKIJ
Dr. N. V. KOTOV
Dr. E. A. GOILO
V. A. SCHITOV
Lehrstuhl für Kristallographie und
Institut für Erdkrustenforschung der
Staatlichen Universität Leningrad
Leningrad/UdSSR, Universitetskaja Naber. 7/9

Kristall und Technik	5	1	1970	145–175

H. J. Bunge

Zentralinstitut für Festkörperphysik und Werkstofforschung
der Deutschen Akademie der Wissenschaften zu Berlin

Some Applications of the Taylor Theory of Polycrystal Plasticity

The Taylor theory of polycrystal plasticity was applied to three-axial deformation accomplished by glide on (111) [110] or (110) [111] glide systems. The orientation dependence of the Taylor factor was used to calculate the angular dependence of the relative strength of a textured material. The angular dependence of the strain ratio R – calculated from the minimum value of the Taylor factor – was compared with the measured strain ratio. The orientation changes of the crystallites of a polycrystalline aggregate after 1% plastic deformation were calculated and compared with experimental values. The orientations which are deformed without orientation changes were calculated and compared with those of maximum orientation density found by a three-dimensional texture analysis of cold rolled copper and iron. The dependence of the rotationless orientations on the axis ratio of the deformation tensor was calculated. This allows suggestions to be made on the dependence of the rolling textures on the lateral broadening of the sheet during rolling.

Die Taylor-Theorie der Vielkristallplastizität wurde auf dreiachsige Verformungszustände und Gleitung auf (111) [110]- oder (110) [111]-Gleitsystemen angewandt. Aus der Orientierungsabhängigkeit des Taylor-Faktors wurde die Winkelabhängigkeit der relativen Festigkeit eines Texturmaterials berechnet. Die Winkelabhängigkeit des R-Wertes der plastischen Verformung – berechnet aus dem Minimalwert des Taylor-Faktors – wurde mit den gemessenen R-Werten verglichen. Die Orientierungsänderungen der Kristallite eines vielkristallinen Aggregates nach 1% plastischer Verformung wurden berechnet und mit experimentellen Werten verglichen. Diejenigen Orientierungen, die sich ohne Orientierungsänderung verformen, wurden berechnet und mit den Orientierungen maximaler Orientierungsdichte verglichen, die sich aus der dreidimensionalen Texturanalyse kaltgewalzter Kupfer- und Eisenbleche ergaben. Die Abhängigkeit der drehungsfreien Orientierungen vom Achsenverhältnis des Deformationstensors wurde berechnet. Daraus ergeben sich Schlußfolgerungen über die Abhängigkeit der Walztexturen von der relativen Breitung des Bleches während der Walzung.

1. Introduction

If a polycrystalline material undergoes a plastic deformation the individual crystallites cannot deform as free single crystals. They are strongly influenced by their neighbours. It must be assumed, therefore, that polyslip takes place. In order to accomplish a general deformation, five independent slip systems are necessary, as was shown by v. Mises. Among the several different combinations of five slip systems that one will be active, which requires the minimum sum of glide shears i.e. the minimum energy. This was shown by Taylor. If one assumes further on, that each crystal undergoes the same deformation as the specimen as a whole, the active combination of slip systems can be calculated for every orientation of the crystallite with respect to the specimen axes. The sum of glide shears — the Taylor factor — is then a measure of the energy required in order to deform a crystallite of this orientation. If the orientation distribution function or texture of the crystallites in a polycrystalline specimen is known, the mean value of the Taylor factor and hence the mean value of the relative strength can be calculated.

If the deformation consists of an elongation in one direction and two different contractions in the two other orthogonal directions, then the Taylor factor depends on the ratio of the two contractions. In the case of a free tensile deformation the contraction ratio is initially not known. It must be assumed, however, that it takes that value which gives rise to a minimum Taylor factor. Thus the contraction ratio can be calculated from the orientation distribution of the crystallites.

Whereas the macroscopic deformation tensor is assumed to be symmetrical, the deformation tensor composed of shears in the different glide systems is generally not symmetrical. If the latter is to be equated to the macroscopic tensor, the antisymmetrical part must be compensated by a crystal lattice rotation in the opposite sense. There are, however, special orientations for which the antisymmetrical part and hence the lattice rotation becomes zero. If the rotations of neighbouring orientations are directed towards these orientations, they are the stable final positions in texture formation.

2. Mathematical considerations

2.1 The Taylor theory for the three-axial strain case

We specify a coordinate system x_1^s, x_2^s, x_3^s in the specimen which coincides with the principle axes of the macroscopic deformation tensor. It is assumed that elongation takes place in one direction — the x_1^s direction — and contraction in the two other directions. A volume element with the form of a cube with an edge length of unity will then take an edge lengths of $1 + \eta$, $1 - q \cdot \eta$ and $1 - (1 - q) \cdot \eta$ in the x_1^s, x_2^s, x_3^s directions, respectively, after deformation (Figure 1). Thus the deformation tensor is given by equation (1)

$$E_{kl}^s = \eta \cdot \begin{vmatrix} 1 & 0 & 0 \\ 0 & -q & 0 \\ 0 & 0 & -(1-q) \end{vmatrix} \tag{1}$$

It is sufficient to consider contraction ratios q in the range of 0 to 0.5, for any symmetric deformation tensor may be reduced to this range of q by a change of the axes and of the sign of η. The degree of deformation η is a small quantity for each deformation step. For the following calculations it is, however, convenient to put η equal to unity (and to reintroduce it only at the very end).

Now we introduce another coordinate system for each crystal $x_1^c = [100]$, $x_2^c = [010]$, $x_3^c = [001]$. The coordinate transformation is given by the equation

$$x_i^c = a_{ij}\, x_j^s \tag{2}$$

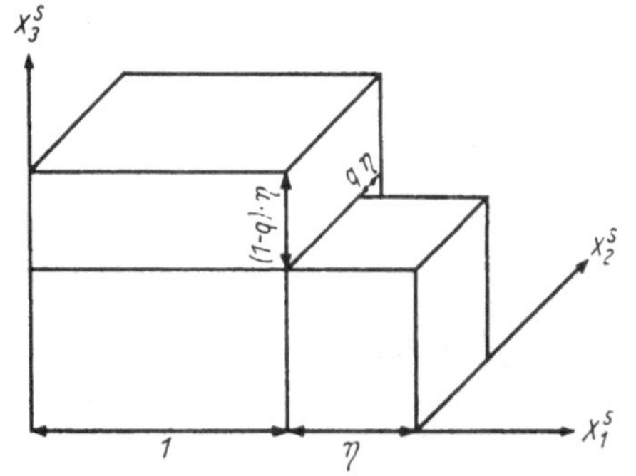

Fig. 1. The volume element before and after deformation

The transformation coefficients a_{ij} are functions of the crystal orientation. They are given in terms of the Euler angles φ_1, Φ, φ_2 by equations (3)

$$\left.\begin{aligned}
a_{11} &= \cos\varphi_1 \cos\varphi_2 - \sin\varphi_1 \sin\varphi_2 \cos\Phi \\
a_{12} &= \cos\varphi_2 \sin\varphi_1 + \sin\varphi_2 \cos\varphi_1 \cos\Phi \\
a_{13} &= \sin\varphi_2 \sin\Phi \\
a_{21} &= -\cos\varphi_1 \sin\varphi_2 - \sin\varphi_1 \cos\varphi_2 \cos\Phi \\
a_{22} &= -\sin\varphi_2 \sin\varphi_1 + \cos\varphi_2 \cos\varphi_1 \cos\Phi \\
a_{23} &= \cos\varphi_2 \sin\Phi \\
a_{31} &= \sin\varphi_1 \sin\Phi \\
a_{32} &= -\cos_1\varphi \sin\Phi \\
a_{33} &= \cos\Phi
\end{aligned}\right\} \tag{3}$$

The components of the deformation tensor in the crystal coordinate system are then given by the transformation relation

$$E^c_{ij} = a_{ik}\, a_{jl}\, E^s_{kl} \tag{4}$$

With equation (1) we can write equation (4)

$$E^c_{ij} = a_{i1}\, a_{j1} - q \cdot a_{i2}\, a_{j2} - (1-q) \cdot a_{i3}\, a_{j3} \tag{5}$$

Now the macroscopic deformation tensor is to be composed of shear components in the different glide systems. For this purpose we introduce a third coordinate system x^g_i for each glide system, the x^g_1-direction being the glide direction, the x^g_3-direction being the normal to the glide plane. In this coordinate system the shear deformation is described by the tensor

$$\varepsilon^g_{kl} = b \cdot \begin{vmatrix} 0 & 0 & 1 \\ 0 & 0 & 0 \\ 0 & 0 & 0 \end{vmatrix} \tag{6}$$

The factor b is the magnitude of the shear in this glide system. The deformation tensor equation (6) may be divided into a symmetrical and an antisymmetrical part

$$\varepsilon^g_{kl} = \frac{1}{2}\, b \begin{vmatrix} 0 & 0 & 1 \\ 0 & 0 & 0 \\ 1 & 0 & 0 \end{vmatrix} + \frac{1}{2}\, b \begin{vmatrix} 0 & 0 & 1 \\ 0 & 0 & 0 \\ -1 & 0 & 0 \end{vmatrix} \tag{7}$$

The latter one describes a rotation of the volume elements of the crystal in the positive sense about the positive x_2^g-direction, the crystal orientation remaining constant, however. The rotation angle β is given by

$$\operatorname{tg} \beta = \frac{1}{2} b \tag{8}$$

The glide direction may be specified by the direction cosines d_i with respect to the crystal axes. The glide plane normal may be specified by the direction cosines p_i. Then the components of the shear tensor, equation (6), with respect to the crystal coordinate system are given by

$$\varepsilon_{ij}^c = b\, p_i\, d_j = b\, e_{ij} \tag{9}$$

If glide takes place on several glide systems the resulting deformation tensor is the sum of the tensors corresponding to the different glide systems

$$\varepsilon_{ij}^c = \sum_n b_n\, e_{ij,n} \tag{10}$$

In cubic face centered crystals the glide planes are (111)-planes and the glide directions [110]-directions. They may be numbered arbitrarily in the way given in Table 1. The coefficients $e_{ij,n}$ are shown in Table 2.

Table 1
The glide systems in f.c.c. crystals

Glide plane	Glide direction		
(111)	1 $[1\bar{1}0]$	2 $[10\bar{1}]$	3 $[01\bar{1}]$
$(\bar{1}11)$	4 [110]	5 [101]	6 $[01\bar{1}]$
$(1\bar{1}1)$	7 [110]	8 $[10\bar{1}]$	9 [011]
$(11\bar{1})$	10 $[1\bar{1}0]$	11 [101]	12 [011]

Table 2
The coefficients $e_{ij,n}$ for the twelve glide systems of f.c.c. crystals given in Table 1

ij \ n	1	2	3	4	5	6	7	8	9	10	11	12	
11	1	1	0	$\bar{1}$	$\bar{1}$	0	1	1	0	1	1	0	
12	1	1	0	1	1	0	$\bar{1}$	$\bar{1}$	0	1	1	0	
13	1	1	0	1	1	0	1	1	0	$\bar{1}$	$\bar{1}$	0	
21	$\bar{1}$	0	1	$\bar{1}$	0	$\bar{1}$	1	0	1	$\bar{1}$	0	1	
22	$\bar{1}$	0	1	1	0	1	$\bar{1}$	0	$\bar{1}$	$\bar{1}$	0	1	$\cdot \frac{1}{\sqrt{6}}$
23	$\bar{1}$	0	1	1	0	1	1	0	1	1	0	$\bar{1}$	
31	0	$\bar{1}$	$\bar{1}$	0	$\bar{1}$	1	0	$\bar{1}$	1	0	1	1	
32	0	$\bar{1}$	$\bar{1}$	0	1	$\bar{1}$	0	1	$\bar{1}$	0	1	1	
33	0	$\bar{1}$	$\bar{1}$	0	1	1	0	$\bar{1}$	1	0	$\bar{1}$	$\bar{1}$	

If we put

$$\varepsilon_{ij}^{c'} = \frac{1}{2}(\varepsilon_{ij}^{c} + \varepsilon_{ji}^{c}) \tag{11}$$

$$\varepsilon_{ij}^{c''} = \frac{1}{2}(\varepsilon_{ij}^{c} - \varepsilon_{ji}^{c}) \tag{12}$$

then, according to the assumption of TAYLOR that each crystal undergoes the same deformation as the specimen as a whole, the following equation must be fulfilled:

$$E_{ij}^{c} = \varepsilon_{ij}^{c'} \tag{13}$$

In this case the antisymmetrical part of the deformation tensor ε must be compensated necessarily by a crystal lattice rotation represented by a rotation vector $\vec{r}$ with the components r_k with respect to the crystal axes

$$r_k = -\varepsilon_{ij}^{c''} \tag{14}$$

Equation (13) can be written

$$E_{ij}^{c} = \sum_{n} \frac{1}{2} b_n (e_{ij,n} + e_{ji,n}) \tag{15}$$

Since the tensor E_{ij}^{c} is symmetrical and the sum of the diagonal elements is zero, only five of the nine equations (15) are linearely independent. A solution for the twelve quantities b_n may be chosen therefore in the from

$$b_{n_1} \ldots b_{n_5} \neq 0 \tag{16}$$

$$b_{n_6} \ldots b_{n_{12}} = 0 \tag{17}$$

where the numbers $n_1 \ldots n_5$ represent an arbitrary set of five linearely independent glide systems. The different combinations may be distinguished by the number m. Thus equation (15) gives rise to several sets of five equations with five unknowns

$$E_{(ij)}^{c} = \sum_{\lambda=1}^{5} b_{n_\lambda}^{m} \cdot \frac{1}{2} (e_{(ij),n_\lambda} + e_{(ji),n_\lambda}) \tag{18}$$

the solution of which may be written in the form

$$b_{n_\lambda}^{m} = \sum_{(ij)=1}^{5} E_{(ij)}^{c} h_{(ij),n_\lambda}^{m} \tag{19}$$

According to TAYLOR it is assumed that each glide system requires the same shear stress τ in order to become active. The energy dissipated in the n-th glide system is then given by $\tau \cdot |b_n|$. Hence, the energy required for the combination of five glide systems is given by

$$W^m = \tau \cdot \sum_{\lambda=1}^{5} |b_{n_\lambda}^{m}| \tag{20}$$

Finally, it is assumed that only those combinations can become active the required energy of which and hence the sum of glide shears is a minimum. If we put

$$M = \mathrm{Min} \sum_{\lambda=1}^{5} |b_{n_\lambda}^{m}| \tag{21}$$

then the required energy is given by

$$W = \tau \cdot M \tag{22}$$

If, on the other hand, the macroscopic stress system is a pure tensile stress σ in x_1^s-direction then according to equation (1) with $\eta = 1$ the required energy is also given by

$$W = \sigma \tag{23}$$

Hence the Taylor factor

$$M = \frac{\sigma}{\tau} \tag{24}$$

describes the relative strength of the crystal due to its special orientation.

The minimum value M (equation (21)) may be taken by several different glide combinations m_ν. These are energetically equivalent. Small individual deviations from the postulated deformation mode (equation (13)) may thus decide which of these combinations will really become active. If we consider the mean behaviour of a large number of crystallites all of these combinations will occur with equal probability. The mean value of the glide shear in the n-th glide system is then given by

$$\bar{b}_n = \sum_\nu b_n^{m_\nu} \tag{26}$$

According to equations (10) and (14) the components of the vector $\vec{r}$ describing the crystal lattice rotation (in the general case $\eta \neq 1$) are given by the relation

$$r_k = -\eta \sum_{n=1}^{12} \bar{b}_n \cdot \frac{1}{2} (e_{ij,n} - e_{ji,n}) \tag{26}$$

If the degree of deformation η is small the angle of rotation is given by

$$\omega = \frac{1}{2} \eta \, |\vec{r}| \tag{27}$$

Since we described the crystal lattice orientation by the Eulerian angles $\varphi_1, \Phi, \varphi_2$, it is convenient to express the rotation also in these variables.

$$\left.\begin{aligned} \Delta\varphi_1 &= \frac{\sin\varphi_2}{\sin\Phi} r_1 + \frac{\cos\varphi_2}{\sin\Phi} r_2 \\ \Delta\Phi &= \cos\varphi_2 \, r_1 - \sin\varphi_2 \, r_2 \\ \Delta\varphi_2 &= -\frac{\sin\varphi_2}{\tan\Phi} r_1 - \frac{\cos\varphi_2}{\tan\Phi} r_2 + r_3 \end{aligned}\right\} \tag{28}$$

In the case of body centered cubic crystals the glide direction is [111] whereas several crystallographically different glide planes may occur. Since the (110) plane seems to be the most frequent one it is reasonable to use the (110) [111] glide systems as a first approximation. In this case, according to equation (9), the relation holds

$$e_{ij}^{\mathrm{bcc}} = e_{ji}^{\mathrm{fcc}} \tag{29}$$

Hence the Taylor factor is the same as in the f.c.c. case

$$M^{\text{bcc}} = M^{\text{fcc}} \tag{30}$$

The rotation, however, occurs in the opposite sense:

$$\vec{r}^{\text{bcc}} = -\vec{r}^{\text{fcc}} \tag{31}$$

2.2 Mean values of the Taylor factor

If mean values of the Taylor factor over all the orientations g in a polycrystalline material are considered it is usefull to approximate the Taylor factor $M(g)$ by a series of generalized spherical harmonics of appropriate symmetry (Bunge 1965, a)

$$M^L(g) = \sum_{\lambda=0}^{L} \sum_{\mu=1}^{N(\lambda)} \sum_{\nu=1}^{N'(\lambda)} m_\lambda^{\mu\nu} \dot{\dot{T}}_\lambda^{\mu\nu}(g) \tag{32}$$

The coefficients $m_\lambda^{\mu\nu}$ are chosen in such a way, that the mean quadratic deviation at the points g_i is minimized

$$\sum_i [M(g_i) - M^L(g_i)]^2 = \text{Min} \tag{33}$$

It is assumed that the Taylor factor of a polycrystalline textured material is given by the mean value of the Taylor factors of all the orientations present in the material

$$\overline{M} = \int M(g)\, f(g)\, dg \tag{34}$$

where $f(g)$ is the orientation distribution function of the crystallites. If $f(g)$ is developed into a similar series as equation (32) with the coefficients $C_\lambda^{\mu\nu}$ (Bunge 1965, b) then $\overline{M}$ is given by

$$\overline{M} = \sum_{\lambda=0}^{L} \sum_{\mu=1}^{N(\lambda)} \sum_{\nu=1}^{N'(\lambda)} m_\lambda^{\mu\nu} C_\lambda^{\mu\nu} \frac{1}{2\lambda + 1} \tag{35}$$

Now let us consider the Taylor factor of a specimen cut from the material at an arbitrary orientation g_0. The orientation of the new specimen axes x_i^σ with respect to the original ones x_i^s is then described by the rotation g_0. The orientation distribution function $f'(g)$ of the crystallites referred to the new axes may be expressed in terms of the function $f(g)$, referred to the original ones by the relation

$$f'(g) = f(g \cdot g_0) \tag{36}$$

The Taylor factor $\overline{M}(g_0)$ of a specimen with the orientation g_0 is then given by

$$M(g_0) = \int M(g)\, f(g \cdot g_0)\, dg\ . \tag{37}$$

In analogy to equation 35 $\overline{M}(g_0)$ may be expressed in terms of the coefficients $m_\lambda^{\mu\nu}$ and $C_\lambda^{\mu\nu}$ (Bunge 1969)

$$\overline{M}(g_0) = \sum_{\lambda=0}^{L} \sum_{\mu=1}^{N(\lambda)} \sum_{\nu=1}^{N'(\lambda)} \sum_{\sigma=1}^{N'(\lambda)} m_\lambda^{\mu\nu} C_\lambda^{\mu\sigma} \ddot{T}_\lambda^{\nu\sigma}(g_0) \frac{1}{2\lambda + 1}\,. \tag{38}$$

The functions $\ddot{T}_\lambda^{\nu\sigma}(g_0)$ exhibit orthogonal symmetry in both the coordinate systems (Bunge 1965, a).

2.3 Numerical calculations

A computer programme was written in ALGOL-language to calculate the Taylor factor M, the components r_k of the rotation vector, the absolute value $|\vec{r}|$ and the orientation changes expressed in Eulerian angles $\Delta\varphi_1$, $\Delta\Phi$, $\Delta\varphi_2$. The orientation coordinates φ_1, Φ, φ_2 were varied in 10°-steps from 0° to 90° and the contraction ratio q was varied in steps of 0.1 from 0 to 0.4. (The case $q = 0.5$ was treated with a separate programme). The programme was run on a computer NE 503. For each value of q the computation took about five hours. (Tables of the Taylor factor as well as of the components r_k of the rotation vector are available through the author).

3. Axially symmetrical flow

The contraction ratio $q = 0.5$ corresponds to axially symmetrical flow, the x_1^s-axis being the axis of symmetry. In this case it is convenient to exchange the x_1^s- and x_3^s-directions in equation (1). Then the x_3^s-direction becomes the axis of symmetry and the problem is independent of the angle φ_1. The Taylor factor

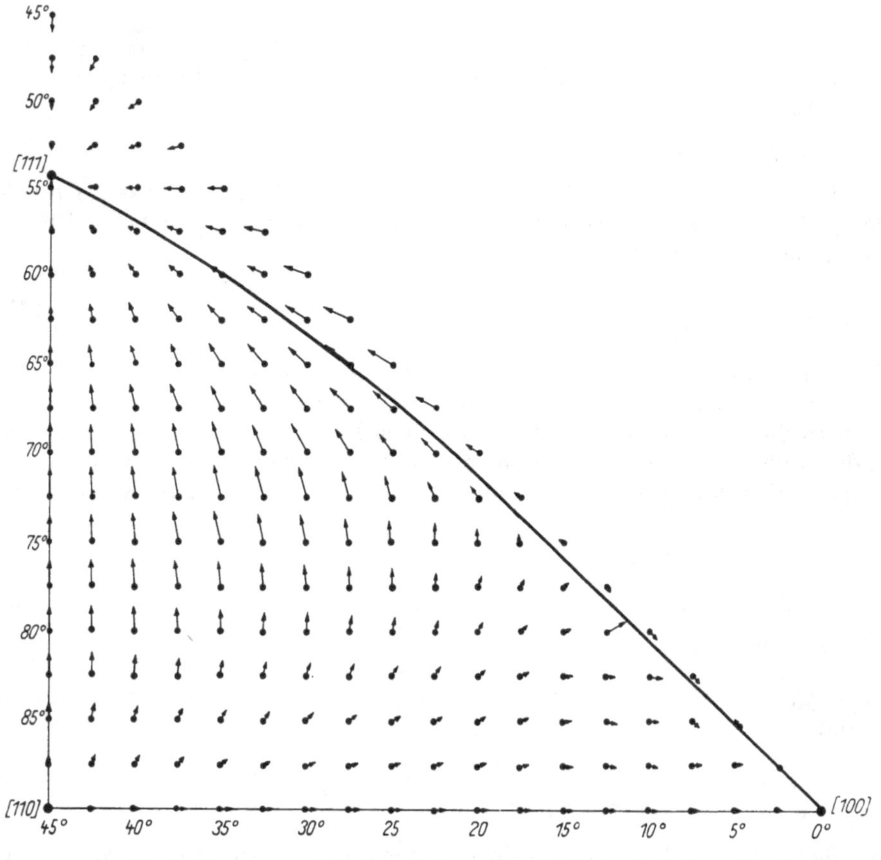

Fig. 2. The mean orientation changes after 3,5 % plastic deformation in the case of axially symmetric flow

calculated in steps of 2.5° for the variables φ_2 and Φ over the whole unit triangle agreed well with the calculations of other authors (CHIN, MAMMEL). It is therefore not shown here.

The orientation change is shown in Figure 2. The end points of the rotation vectors correspond to the orientations attained after 3.5% deformation. They are the mean values of the orientation changes due to the different glide com-

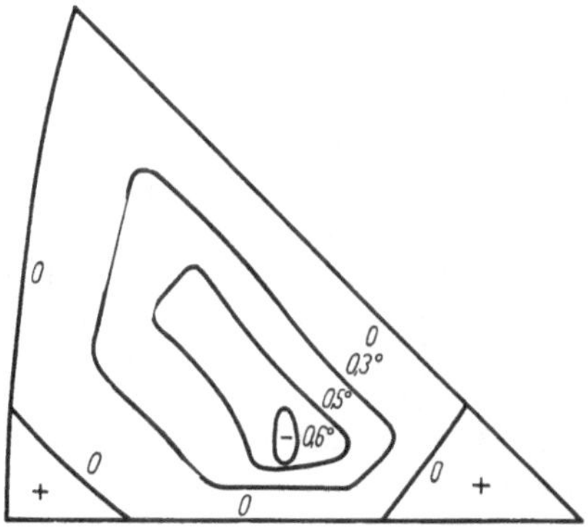

Fig. 3. The mean rotation component about the tensile axis after 3,5 % plastic deformation

binations. It must be mentioned that the rotation law, equation (14), does not agree with that commonly used to describe the orientation changes of freely deformed single crystals. In the latter case the orientation change corresponds to a rotation about an axis perpendicular to the tensile axis and the glide direction. Equation (14), on the other hand, describes rotations about axes perpendicular to the glide directions but in the glide planes. As a consequence of this a rotation component about the tensile axis occurs, the magnitude of which as a function of orientation is shown in Figure 3.

From the rotations shown in Figure 2 the components towards the orientations [111], [100] and [110] were determined. The mean values of these components taken over all orientations with the same angular distance from [111], [100] and [110], respectively, are shown in Figure 4 (corresponding to 1% plastic deformation). They can immediately be compared with the mean orientation chan-

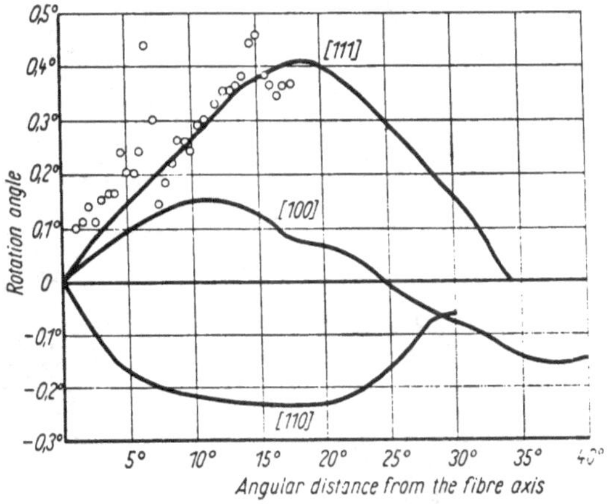

Fig. 4. The orientation changes toward [111], [100] and [110] after 1 % plastic deformation. Measured points after BUNGE and FUCHS

ges towards the tensile direction of the [111], [200], [220] poles of the crystals in a polycrystalline aggregate after 1% plastic deformation. These changes measured for the [111] axes of plastically deformed aluminium wires are shown as points in Figure 4 (BUNGE, FUCHS). The agreement with the theoretical curve is rather good.

4. Plane strain

The plane strain case $q = 0$ corresponds to the ideal rolling deformation without lateral broadening, x_1^s being the rolling direction and x_3^s the normal direction. The deformation tensor consists of an elongation component in the rolling direction and a compression component of the same magnitude perpendicular to the rolling plane. The values of the Taylor factor are available through the author. Values others than those of the 10°-lattice were linearely interpolated and the curves of constant Taylor factor were drawn (Figure 5). This is not quite correct, for the Taylor factor is not a smooth function. It is, however, sufficient if mean values of the Taylor factor for polycrystalline textured materials are to be considered. The maximum value of $2 \cdot \sqrt{6} = 4.90$ occurs in the orientation

Fig. 5. The Taylor factor in the plane strain case $q = 0$ as a function of orientation

Table 3

The coefficients $m_\lambda^{\mu\nu}$ of the series approximating the Taylor factor for different contraction ratios. The error of the coefficients m_0^{11} is about ± 0.0037, that of the other ones about ± 0.012

λ	μ	ν	Plane strain $q = 0$	$q = 0.1$	$q = 0.2$	$q = 0.3$	$q = 0.35$	$q = 0.4$	$q = 0.45$	Radial symm. flow, rolling direction $q = 0.5$	Radial symm. flow, normal direction $q = 0.5$
0	1	1	3.3198	3.2079	3.1415	3.1016	3.0873	3.0766	3.0690	3.0766	3.0766
4	1	1	−1.0861	−1.0026	−0.8899	−0.7461	−0.6694	−0.5913	−0.5105	−0.4189	−1.1167
4	1	2	1.4435	1.3083	1.1282	0.9511	0.8617	0.7726	0.6873	0.6244	
4	1	3	0.1771	0.0394	−0.1563	−0.3701	−0.4843	−0.5979	−0.7099	−0.8257	
6	1	1	−0.0874	−0.0037	0.0431	0.0703	0.0825	0.0911	0.0963	0.0938	−0.3006
6	1	2	−0.0767	−0.1112	−0.1234	−0.1350	−0.1404	−0.1404	−0.1528	−0.1359	
6	1	3	−0.0265	0.0321	0.0963	0.1325	0.1497	0.1641	0.1766	0.1492	
6	1	4	−0.1112	−0.1602	−0.2026	−0.2224	−0.2280	−0.2327	−0.2349	−0.2018	
8	1	1	0.0475	0.0042	0.0191	0.0566	0.0767	0.0946	0.1085	0.1014	0.3802
8	1	2	0.0465	0.0049	−0.0186	−0.0627	−0.0879	−0.1115	−0.1293	−0.1492	
8	1	3	−0.0737	−0.0108	0.0487	0.0950	0.1141	0.1301	0.1438	0.1565	
8	1	4	−0.1413	−0.1528	−0.1587	−0.1729	−0.1729	−0.1683	−0.1661	−0.1742	
8	1	5	0.0269	0.2146	0.3182	0.3243	0.3160	0.3018	0.2814	0.2383	
10	1	1	0.0130	−0.0029	0.0255	0.0421	0.0421	0.0419	0.0382	0.0348	−0.1418
10	1	2	0.0147	−0.0132	−0.0328	−0.0436	−0.0424	−0.0407	−0.0426	−0.0497	
10	1	3	0.0304	0.4083	0.0642	0.0703	0.0659	0.0637	0.0622	0.0512	
10	1	4	0.0181	0.0105	0.0100	−0.0108	−0.0167	−0.0211	−0.0243	−0.0544	
10	1	5	0.0091	−0.0311	−0.0443	−0.0171	−0.0017	0.0093	0.0147	0.0610	
10	1	6	0.0316	0.1019	0.0801	0.0171	−0.0152	−0.0407	−0.0585	−0.0840	

$\varphi_1 = 0$, $\Phi = 90°$, $\varphi_2 = 45°$, this is the $(1\bar{1}0)$ [110] orientation. The minimum value of about $0{,}87 \cdot \sqrt{6} = 2.12$ is taken in the orientation $\varphi_1 = 20°$, $\Phi = 20°$, $\varphi_2 = 10°$. Thus the relative strength of crystals of different orientations varies by about the factor 2.3. In Figure 6 the variation of the Taylor factor is drawn for the orientations $(h\,h\,l)$ $[1\bar{1}0]$ and $(h\,k\,0)$ $[k\,\bar{h}\,0]$. Table 3 gives the coefficients $m_\lambda^{\mu\nu}$ of the series development of the Taylor factor (along with those corresponding to values of q other than $q = 0$).

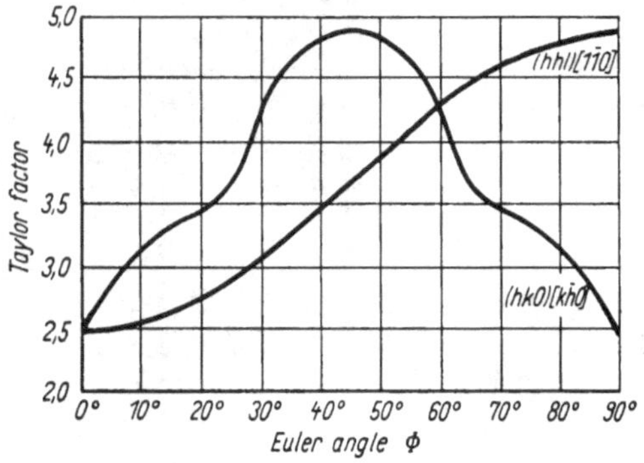

Fig. 6. The Taylor factor for the orientations (h h l) $[1\bar{1}0]$ and (h k 0) $[k\,\bar{h}\,0]$

5. Angular dependence of the rolling resistance

If a sheet material is rolled at the angle α with respect to the original rolling direction the angular dependence of the rolling resistance should be described by the mean Taylor factor $\overline{M}(\alpha)$ according to equation (38), the friction losses being neglected or subtracted. In Figure 7 $\overline{M}(\alpha)$ is given for a temper rolled stabilized steel the orientation distribution of the crystallites of which was determined by the X-ray pole figure technique (BUNGE, ROBERTS). The coefficients

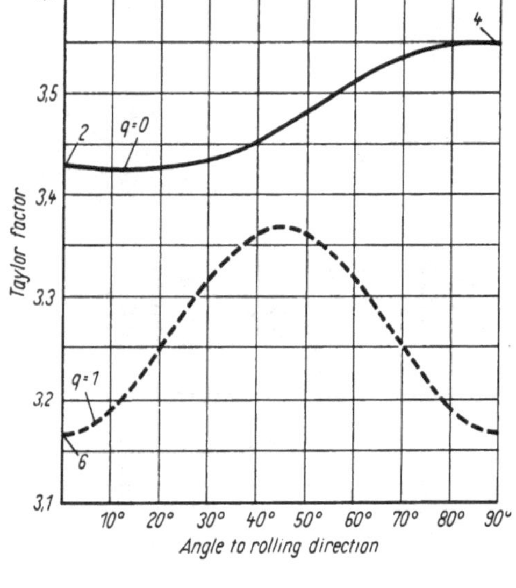

Fig. 7. The Taylor factor for rolling at different angles α to the original rolling direction of a temper rolled stabilized steel ($q = 0$). The dotted curve corresponds to rolling at the angle α but with the rolling plane perpendicular to the original one

$C_\lambda^{\mu\nu}$ are given in Table 4. A comparison with experimental data was, however, not made. The dotted curve in Figure 7 corresponds to rolling deformation with the rolling direction at the angle α to the original rolling direction but with the rolling plane perpendicular to the original one. This curve, corresponding to the contraction ration $q = 1$, will be needed later.

Table 4

The texture coefficients $C_\lambda^{\mu\nu}$ for a temper rolled stabilized steel

ν	$\lambda = 4$	$\lambda = 6$	$\lambda = 8$	$\lambda = 10$
1	−1.4689	2.6861	−0.0707	−1.0248
2	−0.4590	−1.2023	0.2859	0.1910
3	0.4973	0.4641	−0.6541	0.0920
4		−0.1406	−0.4697	−0.2714
5			−0.2219	−0.0488
6				0.0202

6. The ideal orientations of the rolling texture

In Figure 8 the absolute value of the rotation vector $|\vec{r}|$ is drawn as a function of orientation. It exhibits some very sharp minima. Since the values were calculated only for the points of a 10°-lattice of orientations, it is difficult to find the ideal zero positions. These positions can better be found from drawings of the distribution functions of the components r_k of the rotation vector (Figure 9). In Figure 10 the zero lines of these functions in the constant φ_1 sections are shown. They may be stacked one upon another to give an idea of the zero surfaces of the rotation components in the three-dimensional orientation space. A point at which all the three surfaces intersect corresponds to an orientation which deforms without rotation. In Figure 10 several such points are to be seen. Furthermore it turns out, that the zero surfaces run nearly parallel to a certain line through each intersection point. They move away from these lines only moderately. A line element in the orientation space, however, corresponds to a set of orientations related to one another by a rotation about a certain axis. Thus the orientations deforming without rotation correspond to limited fibre axis components as was suggested by Wassermann and Grewen.

The zero orientations may be divided into three classes. If all the neighbouring orientations rotate towards the zero orientation, then it is a stable one. If they rotate away it is an unstable one. If some orientations rotate towards it and other ones away it is an intermediate one. This is illustrated in Figure 11 which shows the Φ- and φ_2-rotation components in the $\varphi_1 = 90°$ section. In the scope of the present theory the stable zero orientations are considered the stable final orientations in the texture formation in the case of f.c.c. metals. According to equation (31) in b.c.c. metals rotation occurs in the opposite sense. Thus the unstable zero orientations of this calculation are to be considered the stable final orientations for b.c.c. metals. As equation (1) shows for the case $q = 0$ a change in sign can also be brought about by an exchange of the rolling direction and the normal to the rolling plane. Hence the rotation vector field for b.c.c. metals is identical to that of f.c.c. metals after the exchange of these two axes. Especially the stable final orientations of the b.c.c. case (the

Fig. 8. The absolute value of the rotation vector as a function of orientation after 3.5 % plastic deformation. The hatched area corresponds to $|\vec{r}| \leqq 0.5°$ —— $r = 1°$ - - - $r = 1.5°$

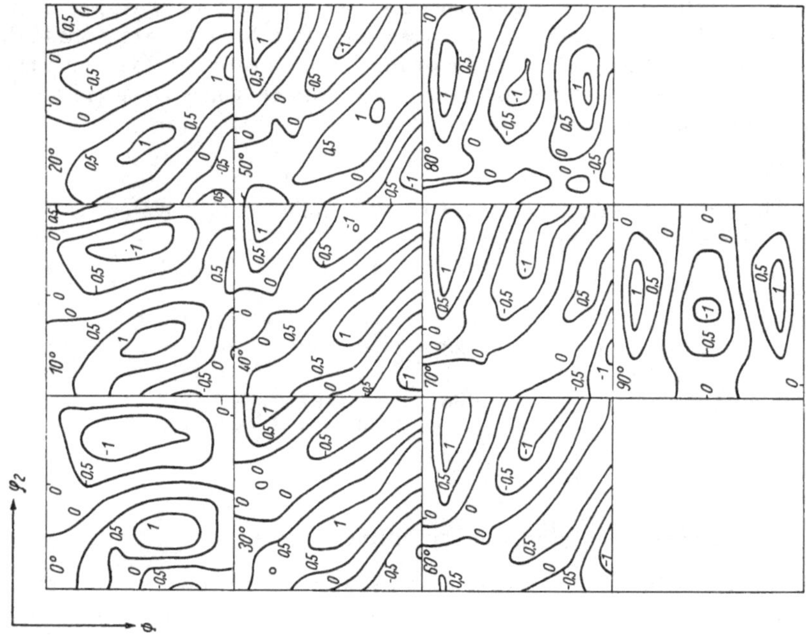

9a

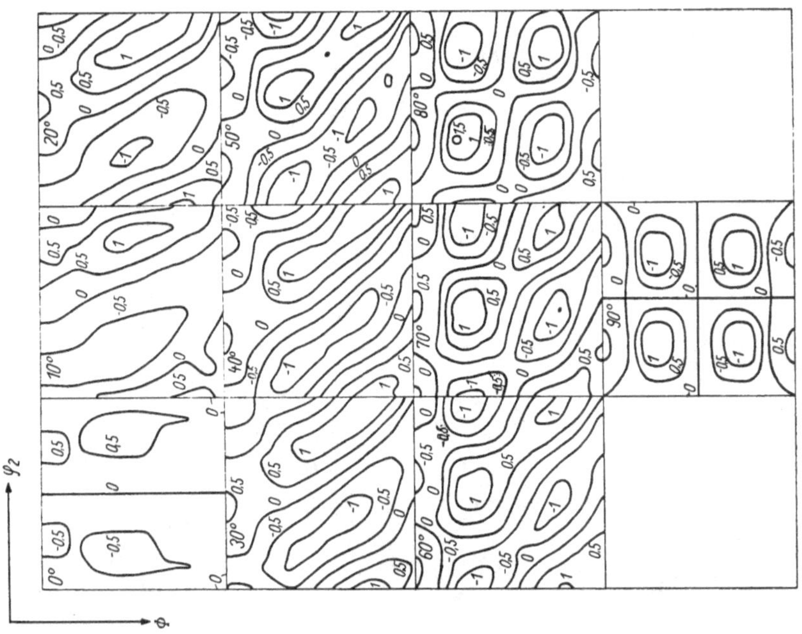

9 c

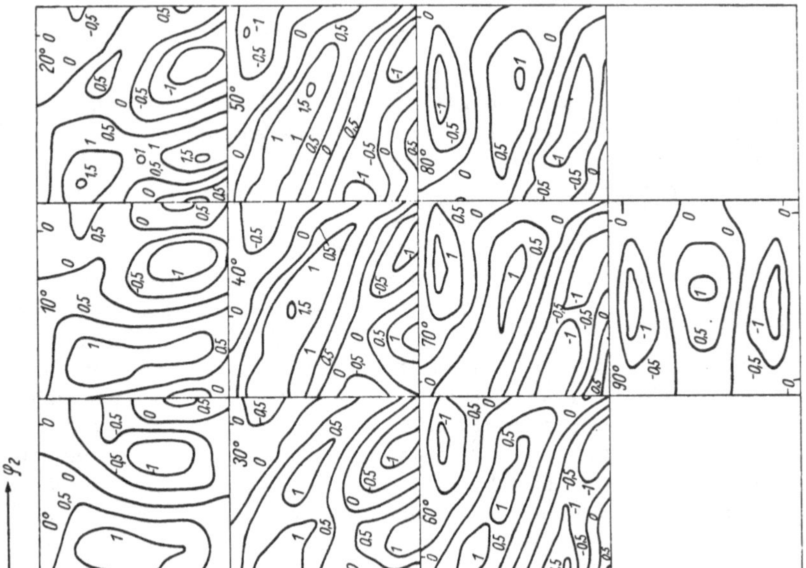

9 b

Fig. 9a–b. The components $r_1(a)$, $r_2(b)$, $r_3(c)$ of the rotation vector in the crystal coordinate system ($x_1^c = [100]$, $x_2^c = [010]$, $x_3^c = [001]$) for different orientations $\varphi_1\ \Phi\ \varphi_2$ in the plain strain case $q = 0$

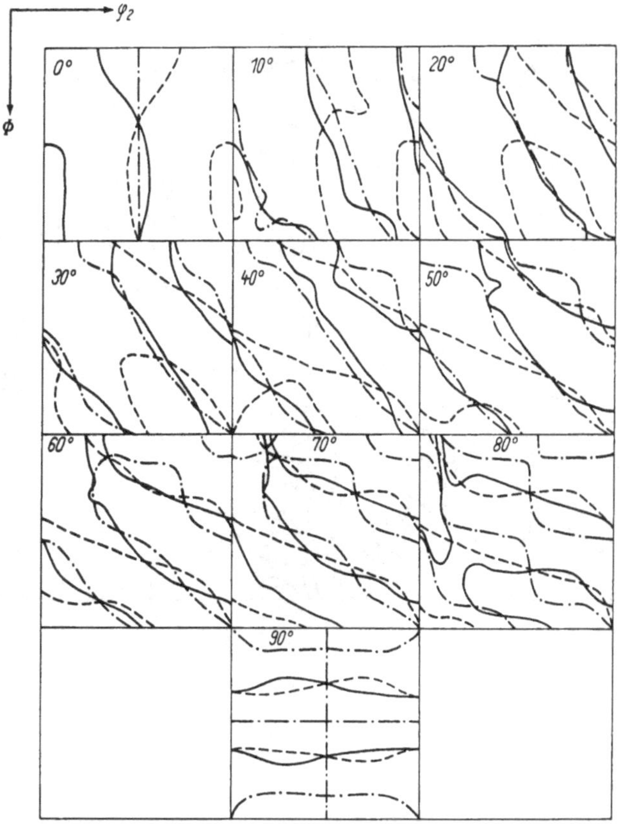

Fig. 10. The zero lines of the components of the rotation vector

instable zero orientations of this calculation) must be identical with those of the f.c.c. case (stable orientations of this calculation) after exchanging rolling direction and normal direction. Hence, the zero orientations must occur in pairs related to one another by an exchange of rolling and normal directions, one of which being stable in the f.c.c. case the other one in the b.c.c. case.

The results of Figure 10 are summarized in Table 5. In the second column the Euler angles of the orientations of exact intersection of all three zero surfaces are given (underlined) as well as those corresponding to the nearly intersection lines, the third column giving in this case the mean deviation of the zero surfaces from these lines. In the fourth column the indices of the underlined orientations are given. The fifth column indicates the rotation axis by which the orientations of one group are related to one another (limited fibre axis). Each group is denoted by a symbol (first column). An asterisk indicates the exchange of rolling and normal direction. The sixth and seventh columns indicate the stable zero orientations, while the last one gives the approximate volume fraction of orientations rotating towards this orientation. In Figure 12 the four stable components of the f.c.c. case are shown by the orientations of rolling and normal directions (heavy lines) in sterographic projection. It is to be seen that the Cu-component does not pass through the highly symmetrical $(1\bar{1}2)$ $[\bar{1}11]$ — position

Table 5

The orientations with zero and nearly zero rotation in the ideal plane strain case $q = 0$

Symbol	Euler angles φ_1	Φ	φ_2	Deviat.	Indices	Fibre axis	Stability f.c.c.	Stability b.c.c.	Volume fraction
Cu	90	27	45	0°	$(4\,\bar{4}\,11)\,[\overline{11}\,11\,8]$		S		
	80	27.5	52	0°	$\sim(1\bar{1}2)\,[\bar{1}11]+8°$				
	70	28.5	58.5	0.7°		RD (45°)ND			
	60	29.5	65	1°					strong
	50	31	71	2°					
Cu*	90	63	45	0°	$(\overline{11}\,11\,8)\,[4\,\bar{4}\,11]$	RD (45°) ND		S	
	80	64	53.5	0.7°	$\sim(\bar{1}11)\,[1\bar{1}2]+8°$				
	70	67	63	1.5°					
Br	35	45	0	0°	$(101)\,[\bar{1}21]$	RD	S		
	32	51	5	2°					
Br*	0	35	45	0°	$(\bar{1}21)\,[101]$	ND		S	strong
	5	35	45	2°					
G	0	45	0	0°	(101) [010]	ND	S		
	5	45	0	1°					medium
G*	45	90	90	0°	(010) [101]	RD		S	
	45	85	85	1°					
[Q]	0	90	0	0°	$(hk0)\,[k\bar{h}0]$	TD			
	0	90	90	0°					
Q	0	90	45	0°	$(101)\,[\bar{1}01]$	RD			
	0	85	45	0.3°					
	0	80	45	0.5°					
	0	70	45	2°					
Q*	0	90	45	0°	$(101)\,[\bar{1}01]$	ND			
	5	90	45	0.3°					
	10	90	45	0.5°					
	20	90	45	2°					
OC	45	12	45	2°	$(1\bar{1}5)\,[051]$	RD (45°) ND	S		
OC*	78	12	0	2°	$(051)\,[1\bar{1}5]$	RD (45°) ND		S	weak

as was frequently expected. Its intersection with the (110)-mirror plane occurs about 8° away from this position at about (4, $\bar{4}$, 11) [$\overline{11}$, 11, 8]. The same stable orientation was recently found theoretically by Dillamore, Buttler and Green. The Br-component is the well known brass orientation (101) [$\bar{1}$21] with a spread about the [$\bar{1}$21] — rolling direction. The Goss-orientation G is spread about the (101)-normal direction. The "off cube" orientation OC, finally, is slightly spread about an axis at 45° to the rolling and normal directions.

Figure 13 shows the "skeleton line" (the line of maximum orientation density) of the orientation distribution of the crystallites in 90% cold rolled copper sheet

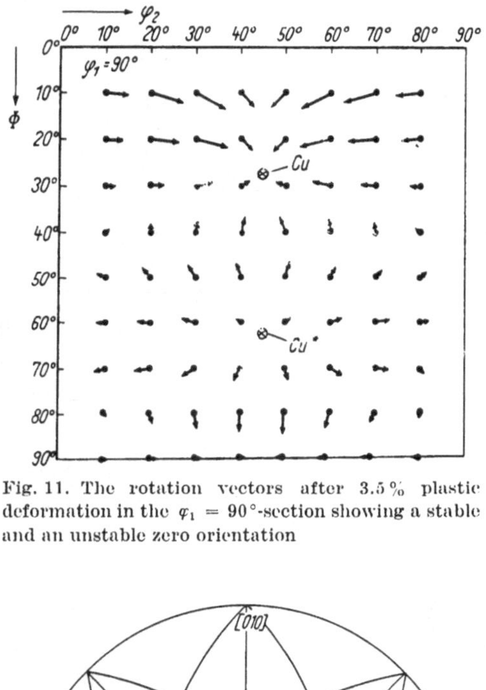

Fig. 11. The rotation vectors after 3.5% plastic deformation in the $\varphi_1 = 90°$-section showing a stable and an unstable zero orientation

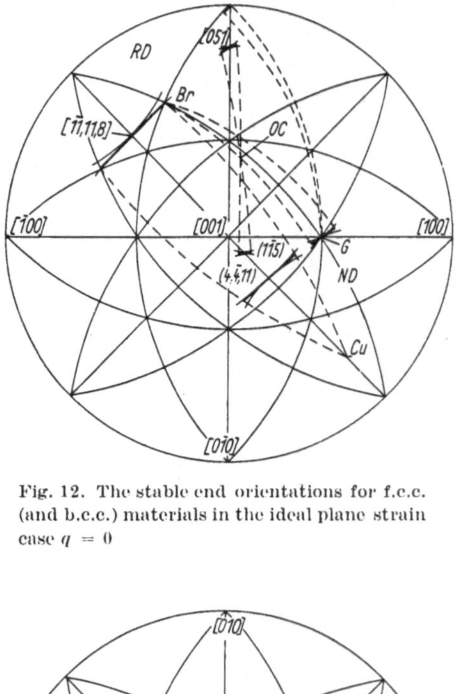

Fig. 12. The stable end orientations for f.c.c. (and b.c.c.) materials in the ideal plane strain case $q = 0$

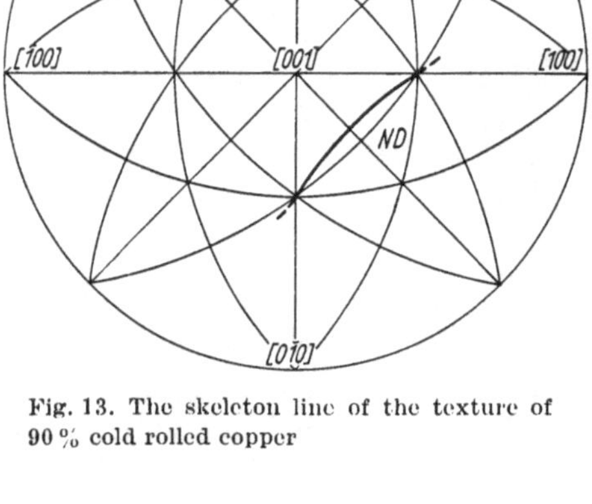

Fig. 13. The skeleton line of the texture of 90% cold rolled copper

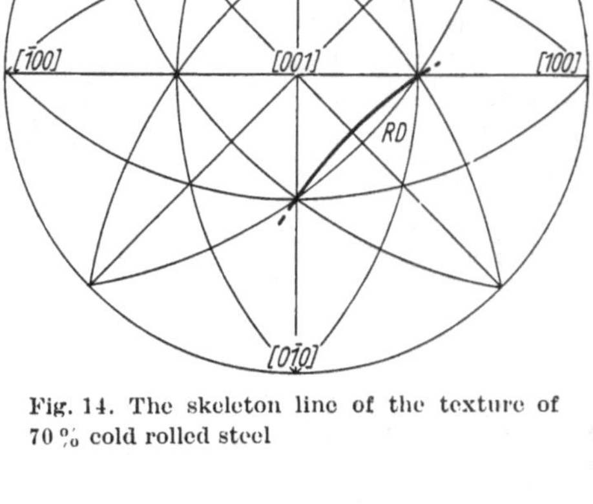

Fig. 14. The skeleton line of the texture of 70% cold rolled steel

as determined by neutron diffraction by BUNGE and TOBISCH. It is very similar to the two main components, Cu and Br, of Figure 12. The third stable component G was not found. There is, however, a faint component only scarcely beyond the limit of experimental error in an orientation near the weak component OC. Figure 14 shows the skeleton of the texture of 70% cold rolled steel (BUNGE 1968). In this case the two main components Cu* and Br* and the median component G* occur. The weak one was not detected. In Figure 15 the orientation distribution function of the copper texture mentioned above is compared with the ranges of low rotation, deduced from Figure 8, corresponding to the two main components Cu and Br. The deviations are not very large. In Figure 16 the same is shown for the steel texture. Here also the third texture component G* is drawn.

Thus, in both cases, the f.c.c. case and the b.c.c. case, the two main components, deduced from this theory, agree fairly well with the experimentally determined textures. It is, however, not understood why the median component does occur in the b.c.c. case but it does not do so in the f.c.c. case. The same

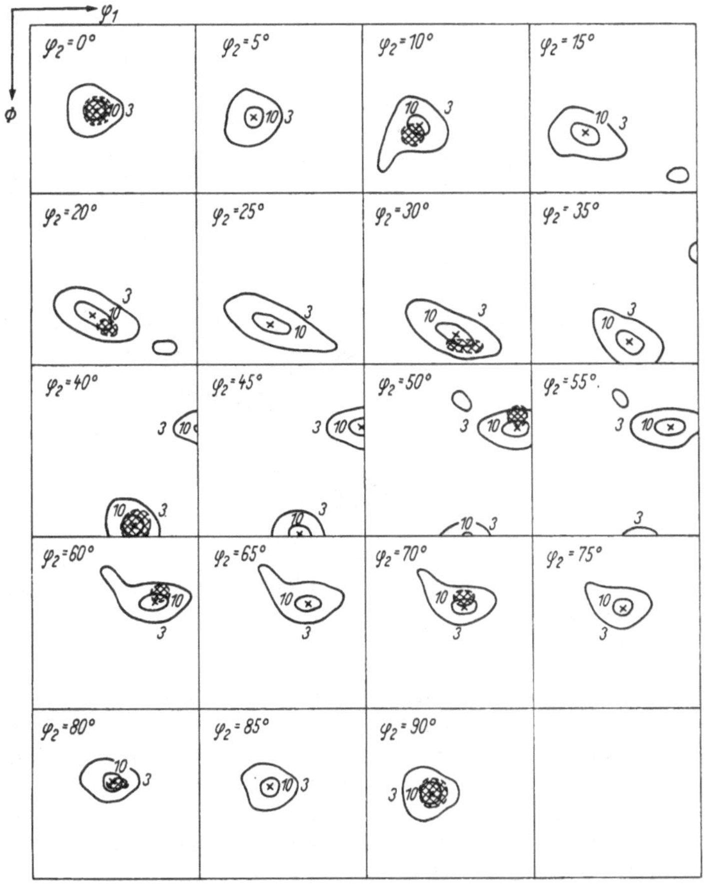

Fig. 15. The three-dimensional orientation distribution function of 90% cold rolled copper sheet compared with the ranges of small rotation vector

holds for the differences in the spread of orientations about the lines of maximum orientation density. Thus the assumptions made in the present calculations seem to be suitable to understand the main features of the rolling textures of f.c.c. and b.c.c. metals. They are, however, not suitable to explain all the details of the textures. In order to understand the deformation textures completely

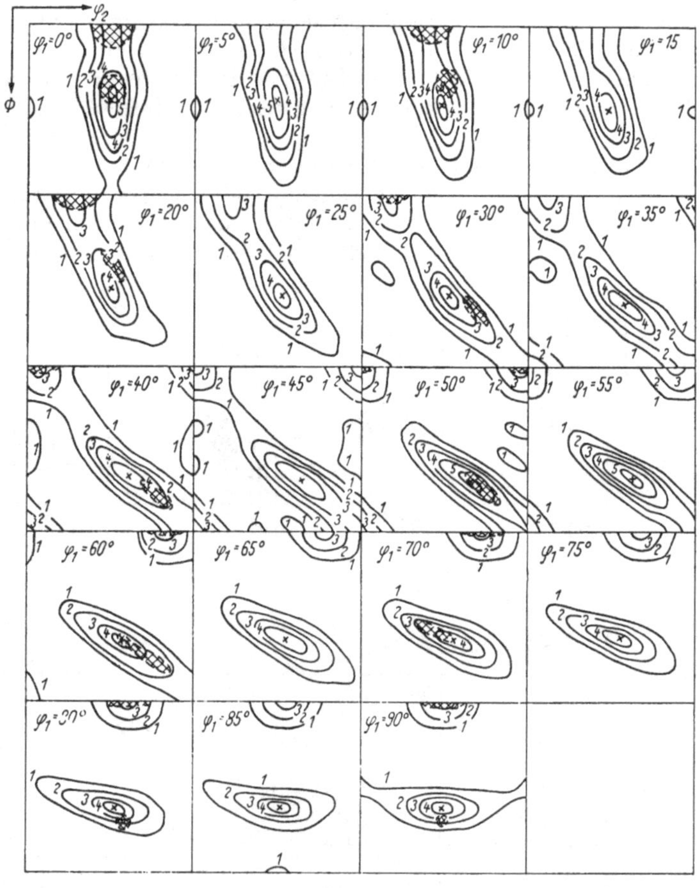

Fig. 16. The three-dimensional orientation distribution function of a 70 % cold rolled steel

further influences, such as other than (110) glide planes in the b.c.c. case, deformation twinning or the inhomogeneity of deformation have to be taken into account. It is, however, also necessary to determine the orientation distribution functions experimentally at an improved degree of resoultion in order to make sure whether a weak component really occurs or not. One of the restricting assumptions involved in the calculations of the stable orientations, the approximation of the rolling deformation by a plane deformation tensor $q = 0$, is abandoned later on. This leads to different changes of the zero orientations in the f.c.c. and b.c.c. case, respectively.

7. Three-axial strain

The Taylor factors for the three-axial strain cases $q = 0.1$; 0.2; 0.3; 0.4 were also calculated. They are given in the Figures 17 a – d. The dependence of the Taylor factor on the contraction ratio q for some special orientations and the random texture is shown in Figure 18. The values corresponding to contraction ratios $q > 0.5$ can be deduced from those for $q < 0.5$ by a rotation of 90° about the x-axis of the specimen. In Table 3 the coefficients $m_\lambda^{\mu\nu}$ of the series developments of the Taylor factors are given for the case $L = 10$ for different contraction ratios. The results of two further calculations of the Taylor factor for the contraction ratios $q = 0.35$ and $q = 0.45$ are included here. The coefficients in the last column correspond to axial symmetric flow with the axis of symmetry parallel to the sheet normal (x_3^s-axis). The zero order coefficient m_0^{11} is the Taylor factor of the ideally random specimen. Its q-dependence is shown in Figure 19.

8. The strain ratio R

If a strip cut at the angle α to the rolling direction of a polycrystalline sheet is deformed in a simple tensile experiment the contraction ratio q is not fixed by the experimental conditions. It must, however, be assumed that it takes on that value $q_{\min}$ for which the Taylor factor $\overline{M}$ is a minimum (Hosford, Backofen)

$$\overline{M}(q_{\min}) = \text{Min} . \tag{39}$$

In many cases the strain ratio

$$R = \frac{q}{1 - q} \tag{40}$$

is used rather than the contraction ratio q. Figure 20 shows the mean Taylor factor $\overline{M}$ calculated from equation (35) with the coefficients $C_\lambda^{\mu\nu}$ of Table 4 as a function of the contraction ratio q for a temper rolled stabilized steel in the rolling direction and at 45°. The minimum of the curves is very flat. Thus small changes of the curves due to small texture changes can strongly influence the exact position of the minimum. This seems to be the reason for the fact, that the strain ratio R or the contraction ratio q are extremely sensitive on small changes of the texture. In Figure 21 the minimum values $q_{\min}$ are drawn as a function of the angle α to the rolling direction. The agreement with the experimentally determined curve is good if one keeps the flatness of the minimum in Figure 20 in mind. The deviations do not exceed the limit given by the errors of the coefficients $m_\lambda^{\mu\nu}$ and the texture coefficients $C_\lambda^{\mu\nu}$.

As a consequence of the errors of these quantities the mean value of the Taylor factor is uncertain by about 0.02, thus giving rise to an uncertainty of about 0.15 in q. It is to be mentioned that an error of 0.02 in the Taylor factor $\overline{M}$ corresponds to less than 1% of the range of variability of the Taylor factor for single crystals in the different orientations.

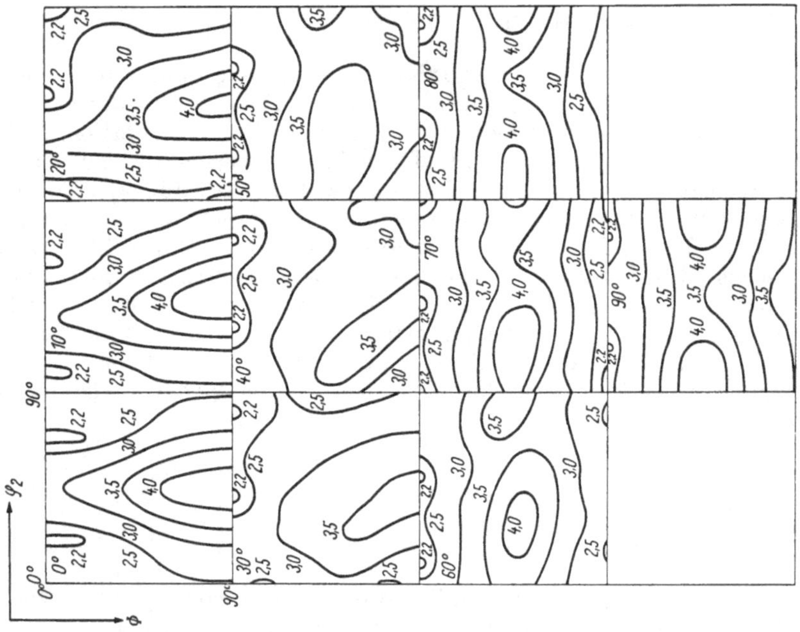
φ2
Φ
0°
90°
0°
10°
20°
30°
40°
50°
60°
70°
80°
90°

17 b

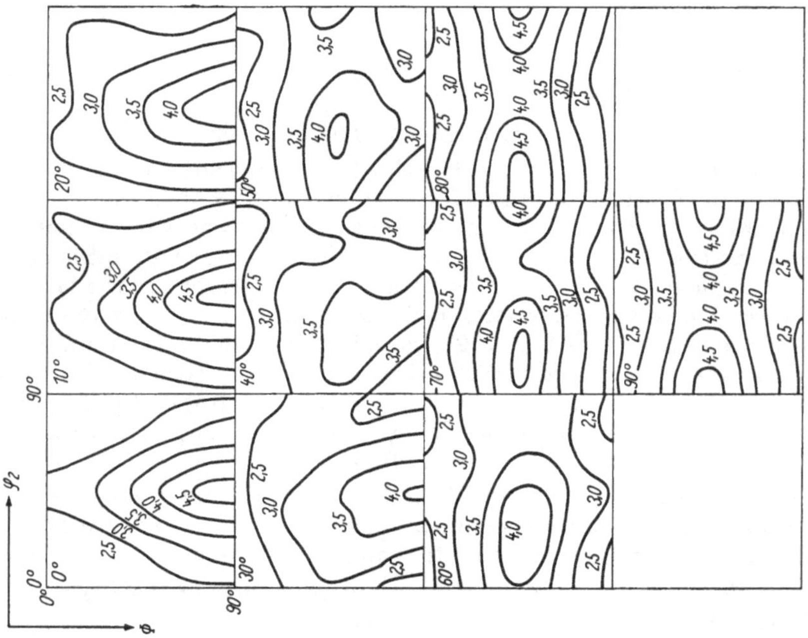
φ2
Φ
0°
90°
0°
10°
20°
30°
40°
50°
60°
70°
80°
90°

17 a

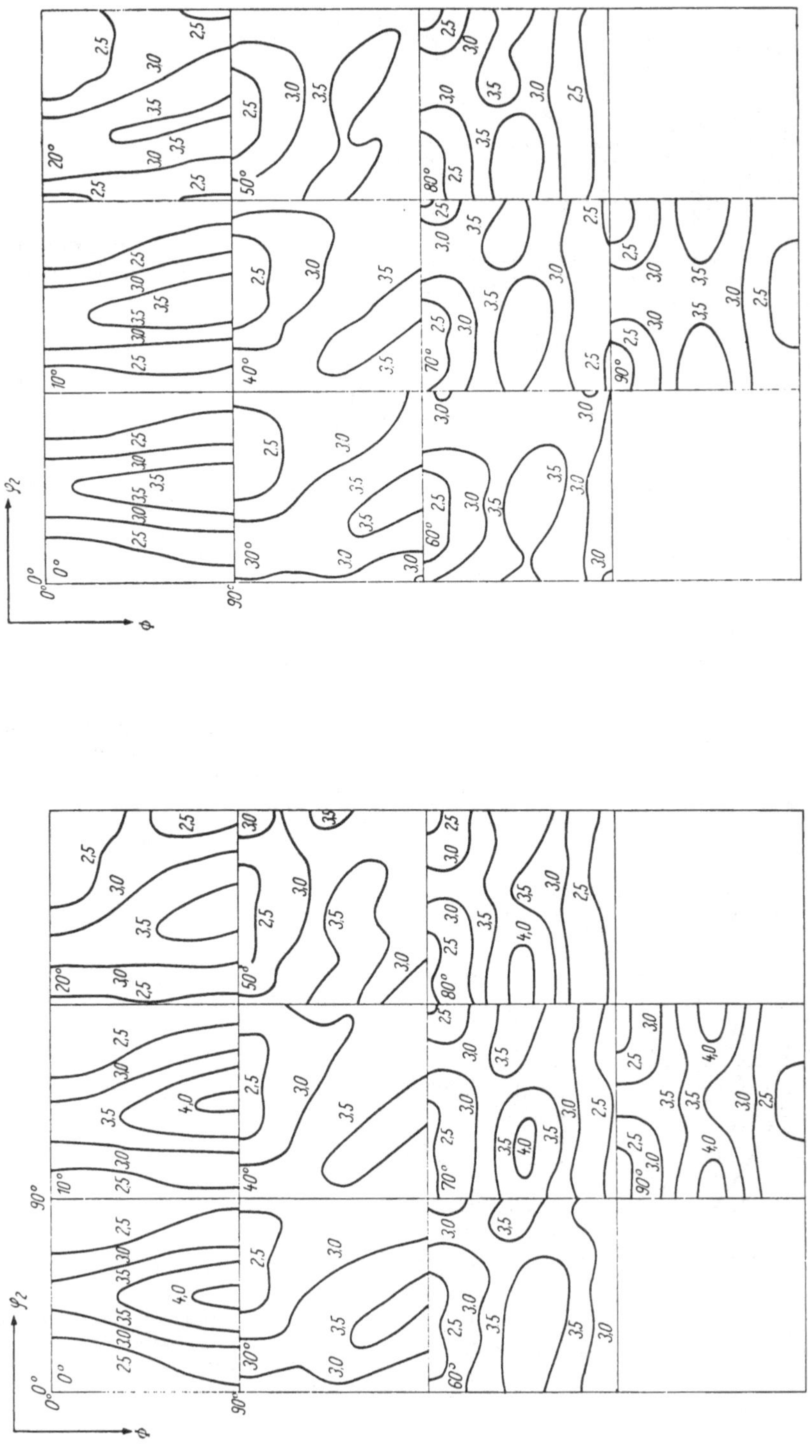

17 c

17 d

Fig. 17. $a-d$. The Taylor factor in the three-axial strain cases $q = 0.1$ (a), $q = 0.2$ (b), $q = 0.3$ (c), $q = 0.4$ (d)

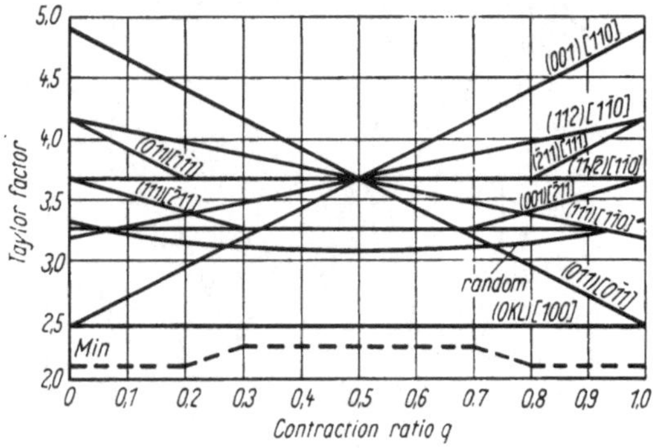

Fig. 18. The Taylor factor of special orientations as a function of the contraction ratio q

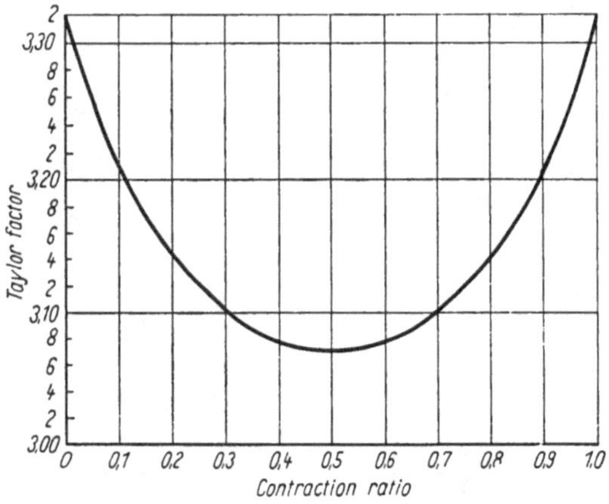

Fig. 19. The Taylor factor for the ideally random specimen as a function of the contraction ratio q

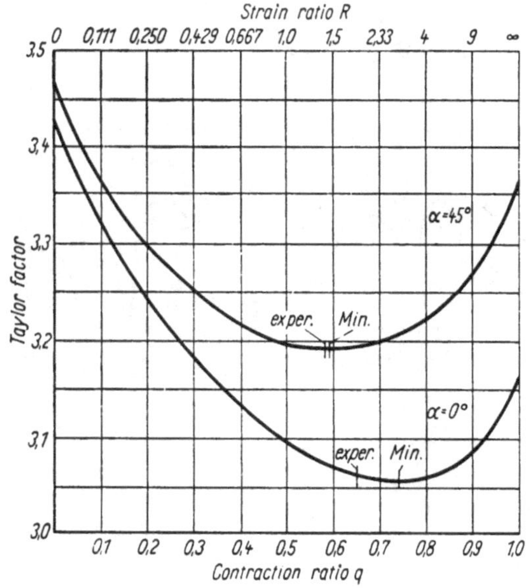

Fig. 20. The mean Taylor factor as a function of the contraction ratio q for specimes cut from a stabilized steel sheet in the rolling direction and at 45°

9. Texture strengthening

In Figure 22 the minimum values $\overline{M}(q_{\min})$ of the Taylor factors of the tensile specimens mentioned in the preceeding section are shown. They are compared with the experimentally determined values of the yield stress. An appropriate factor was chosen to correlate the two scales to one another. The deviations do not exceed the limit of experimental error. The Taylor factor for compression along the normal direction of the sheet was calculated with the coefficients of the last column in Table 3 which correspond to axially symmetric flow with respect to the sheet normal. It was found $\overline{M}_3 = 3.2011$.

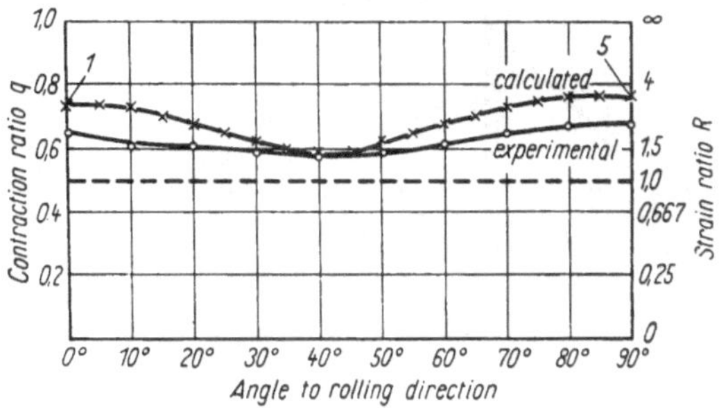

Fig. 21. The contraction ratio q and strain ratio R as a function of the angle α to the rolling direction of a stabilized steel sheet

According to equation (24) the Taylor factor is a measure of the stress needed in order to initiate or maintaine plastic flow in the material. Usually, the stress components necessary to maintain plastic flow in the plane stress case are described by the yield locus. This is the curve in the $\sigma_1 - \sigma_2 -$ plane outside of which plastic flow takes place whereas inside it does not do. As was shown by HOSFORD and BACKOFEN the yield locus can approximately be constructed with the results of plane strain compression tests and free compression and tension tests. The plane strain compression experiments correspond to our calculations for rolling deformation without lateral broadening ($q = 0$ and $q = 1$ in Figure 7). They give the σ_1- and σ_2- and 45°-tangents to the yield locus (points 2, 4 and 6 in Figure 23). The case of free tension in x_2^s direction (Figure 22) gives the intersections with the σ_1- and σ_2-axes, respectively, (points 1 and 5) whereas the q-values of these cases correspond to the slopes of the curve in these points. Finally, the compression in normal direction corresponds to equiaxed stress in the $\sigma_1 - \sigma_2$-plane (point 3). In this way the yield locus, Figure 23,

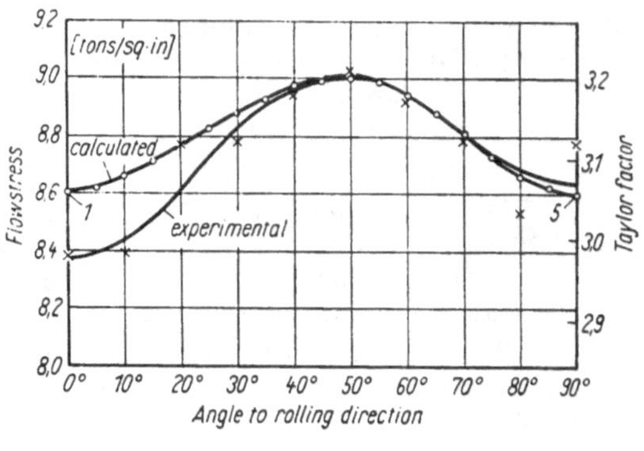

Fig. 22. The dependence of the Taylor factor for $q = q_{\min}$ on the angle α to the rolling direction compared with the experimentally determined flow stress in a stabilized steel heet. (The two scales have been adapted to one another by an appropriately chosen factor)

was constructed. As a consequence of the initial assumptions it is symmetrical with respect to the origin, but it deviates from the elliptical form postulated by HILL for anisotropic materials. The points 1 to 6 used to construct the yield locus are also indicated in the original Figures 7, 21 and 22 from which they were taken. Figure 24 shows the yield locus of the copper texture the three dimensional orientation distribution of which is shown in Figure 15.

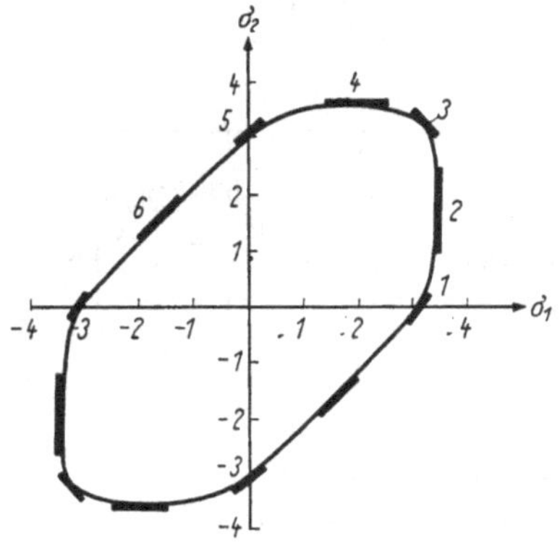

Fig. 23. The yield locus calculated after the Taylor theory for temper rolled stabilized steel

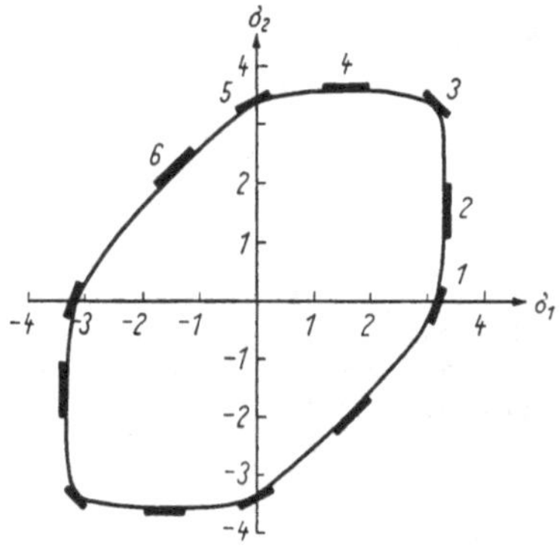

Fig. 24. The yield locus calculated for a 90 % cold rolled copper sheet

10. The dependence of the rolling texture on the relative broadening

In a preceeding section the rolling deformation was approximated by a plane strain model. This means, the lateral broadening of the sheet passing the rolls is not taken into account. In the case of large width to thickness ratios of the sheets this may be a rather good approximation. If the width is small, however, the broadening is generally not negligible. Therefore we now approximate the rolling deformation by a contraction of the magnitude η in the normal direction (x_3^s-direction), an elongation of $(1 - q) \cdot \eta$ in the rolling direction (x_1^s-direction) and an elongation of $q \cdot \eta$ in the transverse direction (x_2^s-direction). The corresponding deformation tensor can be deduced from that of equation (1) by an exchange of rolling and normal directions (x_1^s and x_3^s) and a change in the sign of η.

Figure 25 shows the absolute value of the rotation vector as a function of the orientation and the contraction ratio q (for a deformation tensor given by equation (1)). The change of the positions of the low rotation ranges with respect to the case $q = 0$ is clearly to be seen. The exact zero orientations can, however, better be determined using the components of the rotation vector, as it was already done in the case $q = 0$. In the general three-axial strain case $q \neq 0$ the transition from the f.c.c. to the b.c.c. case, which is equivalent with a change in the sign of η is, however, no longer equivalent with the exchange of the rolling and normal direction. Hence the stable zero orientations (f.c.c.

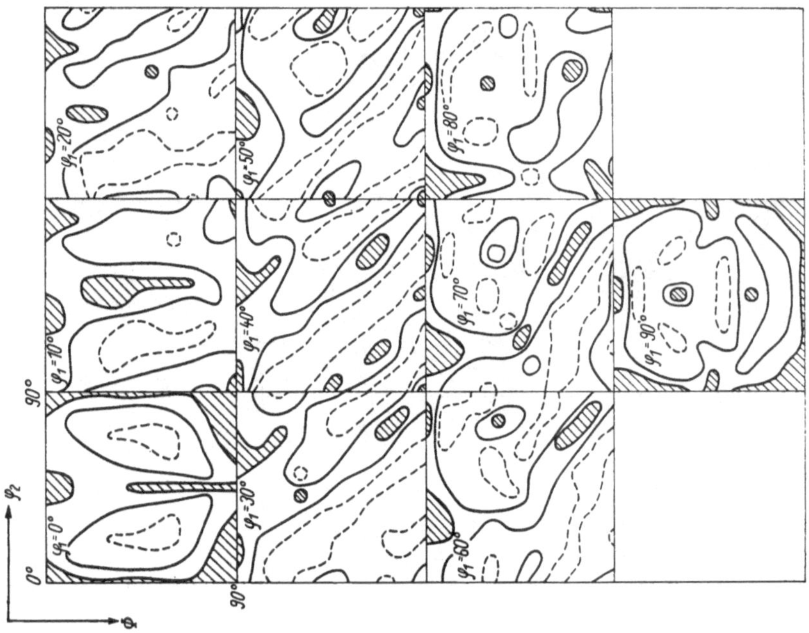
φ2
Φ
0°
90°
90°
φ1=0°
φ1=10°
φ1=20°
φ1=30°
φ1=40°
φ1=50°
φ1=60°
φ1=70°
φ1=80°
φ1=90°

25 b

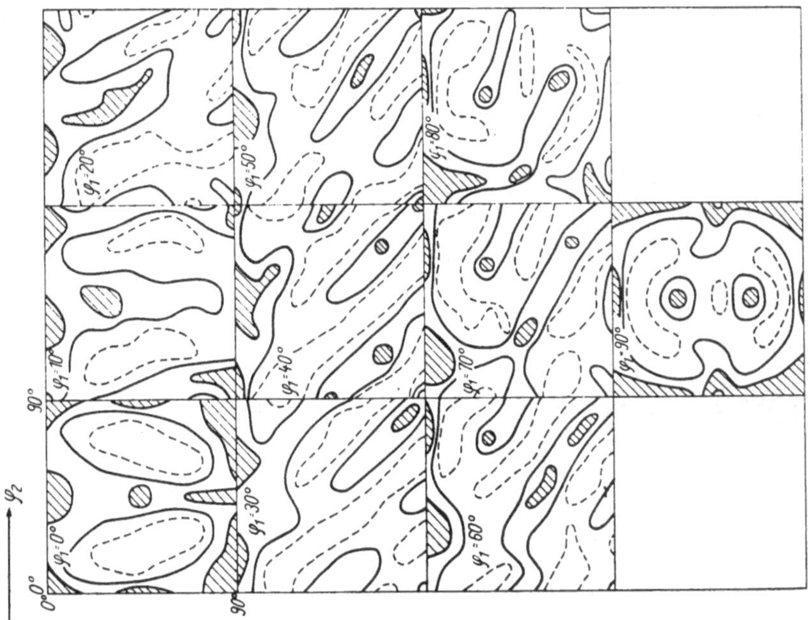
φ2
Φ
0°
0°
90°
90°
φ1=0°
φ1=10°
φ1=20°
φ1=30°
φ1=40°
φ1=50°
φ1=60°
φ1=70°
φ1=80°
φ1=90°

25 a

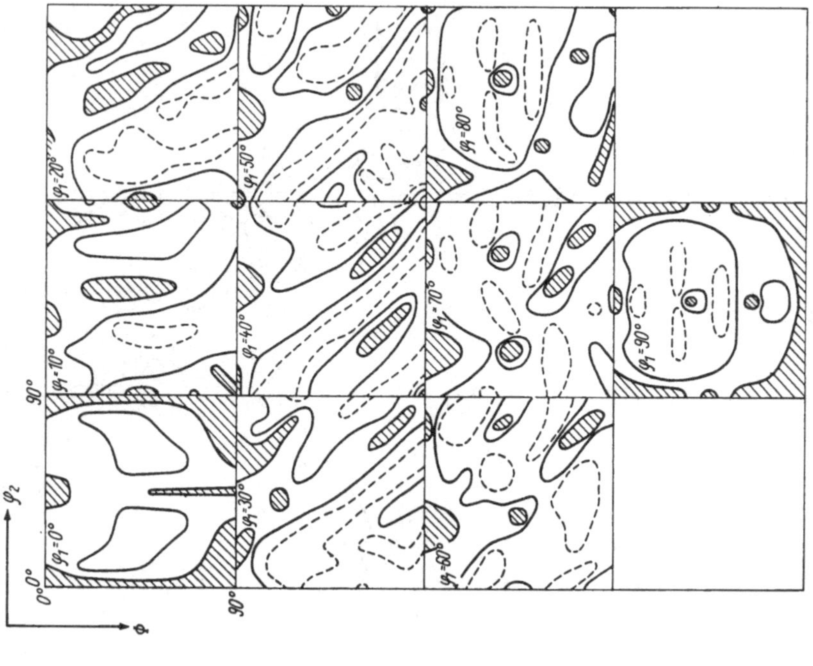

25c

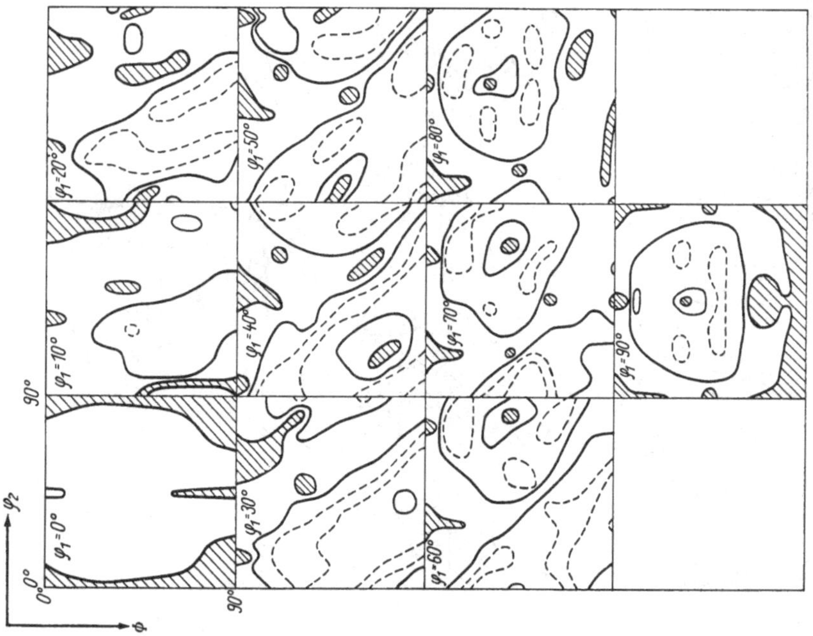

25d

Fig. 25a–d. The absolute value of the rotation vector after 3,5% plastic deformation as a function of orientation for the contraction ratios $q = 0.1$ (*a*); $q = 0.2$ (*b*); $q = 0.3$ (*c*); $q = 0.4$ (*d*) hatched area $|\vec{r}| \leqq 0.5°$, —— $r = 1°$, - - - $r = 1.5°$

orientations) and the unstable ones (b.c.c. orientations) are no longer equivalent. They move away in different directions from the equivalent positions at $q = 0$.

In Figure 26 the f.c.c. case is shown. As q increases the Cu-component moves towards smaller angles δ and disappears at about $q = 0.4$. The Br-component (101) [$\bar{1}$21] moves at first ($q = 0.1$) towards (101) [$\bar{1}$11] and a large spread now occurs about the (101) normal direction. Simultaneously the (101) [$\bar{1}$01)-component Q, which was an intermediate one at $q = 0$, now becomes stable and also a large spread occurs about the (101)-normal direction. At higher values of q these two components merge in one component which, however, leaves the highly symmetric position. At $q = 0.3$ the normal direction deviates about 10° from (101). Finally, however, it moves again towards (101). The (101) [010]-component G extends its scattering angle about the (101)-normal direction. Thus, these three components finally merge in a complete (101)-fibre component about the normal direction, which is the well known stable texture in the case of axially symmetric compression of f.c.c. metals.

In Figure 27 the b.c.c. case is shown. The Cu*- and Br*-components move towards a complete (111)-fibre texture at $q = 0.5$. The (010) [101]-component G* and the weak (051) [1$\bar{1}$5]-component OC* merge in one component and finally yield a complete (010)-fibre texture. The movement of the main components (Cu and Cu*) in different directions away from the common equivalent positions at $q = 0$ is to be seen in Figure 28, which shows the angle δ as a function of q.

Ashown in Figures 26 and 27 the rolling textures ought strongly to be influenced by the relative broadening. This holds especially for the brass orientation (101) [$\bar{1}$21] in f.c.c. metals. There seem to be no systematic experimental investigations of the influence of lateral broadening on the rolling texture. Thus, no comparisons between experimental and theoretical results can be made. The relative broadening depends on many variables such as, for example, the roll diameter, the width to thickness ratio, the roll friction and lubrication,

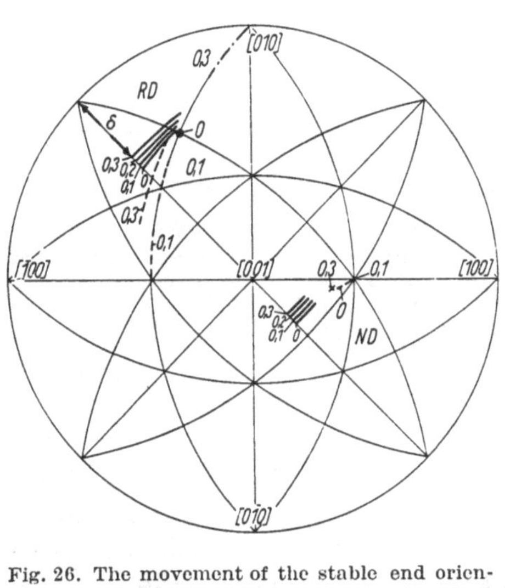

Fig. 26. The movement of the stable end orientations for rolling of f.c.c. metals as a functions of the relative broadening q

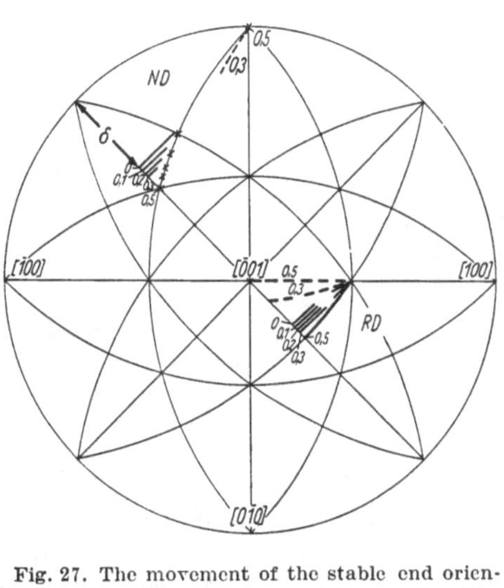

Fig. 27. The movement of the stable end orientations for rolling of b.c.c. metals as a function of the relative broadening q

the material under investigation and many others, too. In most cases these parameters are not known. This may be one of the reasons for differences in the rolling textures found by different investigators. Further more the relative broadening may vary during the deformation process. Hence the "stable zero orientations" themselves move during deformation. They run away from the crystal orientations, which tend to reach them during deformation. This may give rise to a larger amount of spread than would be present with fixed "stable orientations".

The absolute broadening varies only little with the width of the rolled sheet. Thus the relative broadening increases strongly with decreasing strip width, the other parameters remaining constant. This is shown in Figure 29, which was constructed from data published by Lueg and Pomp. Hence a deminution of the strip width ought to bring about the texture changes indicated in Figures 26 and 27. Experiments to realize this effect are under preparation.

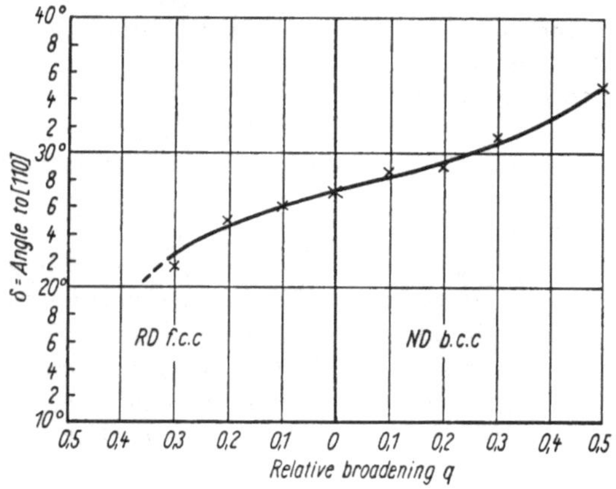

Fig. 28. The angle δ of the main component as a function of the relative broadening q

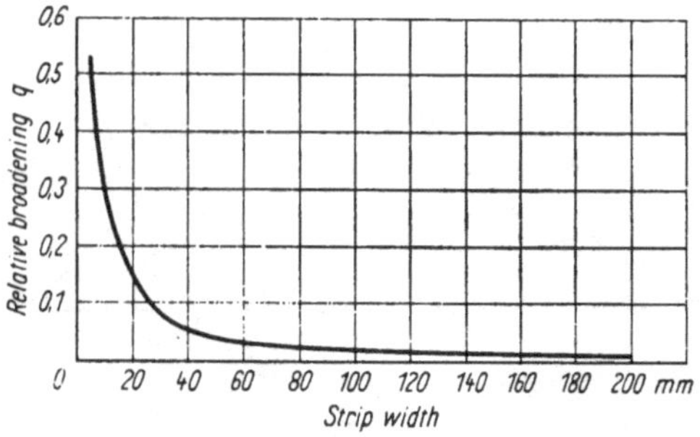

Fig. 29. The relative broadening of sheets after rolling as a function of the strip width, deduced from measurements of Lueg and Pomp

11. The texture in the yield cone of a teared specimen

If a wire is elongated by drawing through dies the flow is radially symmetrical and the well known fibre textures are developed. If it is, however, elongated freely an occasionally present deviation of the radially symmetric distribution of crystal orientations may lead to non radially symmetrical flow as was shown in a preceeding chapter. This flow, in turn, will lead to crystal rotations towards certain stable orientations corrsponding to the actual contraction ratio, different from those for radially symmetrical flow. Now the changed orientation distribution will give rise to a changed contraction ratio and so on. Thus a texture completely different from that in wire drawing may result. It seems, however, rather difficult to calculate the final orientations of this process since the contraction ratio and hence the crystal rotations depend on the actually present texture and both texture and contraction ratio change during the deformation process. The principles of texture formation in this case ought, however, to be understandable in the scope of this theory. Since the contraction ratio is strongly dependent on small differences in the texture, investigations of the texture formation by free tension should offer a good proof of the theory and its basic assumptions.

References

BUNGE, H. J.: Mber. dtsch. Akad. Wiss. **7**, 351 (1965) a
BUNGE, H. J.: Z. Metallkunde **56**, 872 (1965) b
BUNGE, H. J.: Physica Status solidi **26**, 167 (1968)
BUNGE, H. J.: Mathematische Methoden der Texturanalyse, Berlin 1969
BUNGE, H. J., FUCHS, R.: Physica Status solidi **32**, 169 (1969)
BUNGE, H. J., ROBERTS, W. T.: J. appl. Crystallogr. **2**, 116 (1969)
BUNGE, H. J., TOBISCH, J.: Z. Metallkunde **59**, 471 (1968)
CHIN, G. Y., MAMMER, W. L.: Trans. metallurg. Soc. AIME **239**, 1400 (1967)
DILLAMORE, J. L., BUTLER, E., GREEN, D.: Metal sci. J. **2**, 161 (1968)
HILL, R.: Mathematical Theory of Plasticity, London 1950
HOSFORD, W. F., BACKOFEN, W. A.: Strength and Plasticity of Textured Metals in: Fundamentals of Deformation Processing, Syracuse 1964
LUEG, W., POMP, A.: Stahl u. Eisen **53**, 810 (1933)
MISES, R. v.: Z. angew. Math. Mechan. **8**, 161 (1928)
TAYLOR, G. J.: J. Inst. Metals **62**, 307 (1938)
WASSERMANN, G., GREWEN, J.: Texturen metallischer Werkstoffe, Berlin, Göttingen, Heidelberg 1962

(Received March 1, 1969)

Authors address:

Dr. habil. H.-J. BUNGE
Zentralinstitut für Festkörperphysik und Werkstofforschung 27
der Deutschen Akademie der Wissenschaften
zu Berlin
8032 Dresden
Helmholtzstraße 20

Berichtigung

In Heft 4, Bd. 4, 1969 muß es heißen:

S. 441,	Zeile 15 u. 23 v. o.	0,138 anstatt 0138
	Zeile 23 v. o.	*hk0* anstatt *kk0*
	Zeile 3 v. u.	(Abdulgamidov, Fridkin) anstatt Abdulgamidov, Fridkin
S. 442,	Zeile 4 v. o.	(001) anstatt $(00\bar{1})$
	Zeile 12 v. u.	pseudocentering anstatt psendocentering
S. 444,	Zeile 13 v. o.	involves anstatt involved
	Zeile 20 v. u.	steps anstatt Steps
	Zeile 14 u. 15 v. u.	$Z = 8$, $Z = 19$, $Z = 24$ anstatt z = 8, z = 19, z = 24
	Zeile 12 v. u.	mistake) anstatt mistake
S. 445,	Unterschrift Fig. 3.	vector I anstatt vector 1
S. 446,	Zeile 4 v. u.	larger than anstatt equal to $>$
	Zeile 1 v. u.	points $\geqq 8$ anstatt points is $\geqq 8$
S. 448,	Zeile 20 v. u.	fundamental anstatt funfdamental
	Zeile 18 v. u.	point 22 anstatt points 22
	Zeile 9 v. u.	the polygon anstatt their polygons
	Zeile 7 v. u.	$N < 8$ anstatt $N > 8$
S. 449,	Zeile 12 v. o.	copulation anstatt coincidence
	Zeile 2 v. u.	higher anstatt highest
S. 450,	Zeile 20/21 v. o.	matter anstatt substance
S. 450,	Zeile 19 v. u.	step in structure anstatt stop in structur
S. 452,	Zeile 2 v. u.	(Cr_3-tetrahedron), anstatt (Cr_3-tetrahedron,
S. 453,	Zeile 9 v. u.	K_3 anstatt $^{32}K_3$
S. 454,	Zeile 13 v. o.	one and a half anstatt half
	Zeile 23 v. o.	$1{,}5 \times 2 = 3$ anstatt $1{,}5 \times 3 = 3$
	Zeile 2 v. u.	group II = the 2-nd chapter
S. 455,	Zeile 4 v. o.	determined, none the less anstatt determined none the less,
	Zeile 12 v. o.	idea of anstatt idea
	Zeile 16 v. o.	($K_2Cr_2O_7 \rightarrow \cdot\cdot$ anstatt $K_2Cr_2O_4 \rightarrow$
S. 457,	Zeile 1 v. u.	K cation octahedra anstatt K cations – on octahedra
S. 458,	Zeile 17 v. u.	by almost anstatt on almost
	Zeile 16 v. u.	K-bichromate anstatt K-bichrimate
	Zeile 17, 15 v. u.	of two anstatt from two
S. 459,	Zeile 3 v. o.	of K-polyhedra. anstatt from K-polyhedra.
	Zeile 9 v. u.	discrepancy anstatt discrepency
	Zeile 8 v. u.	reflecting anstatt affecting
S. 460,	Zeile 29 v. u.	МИН anstatt МНИ.
	Zeile 25 v. u.	Векторное anstatt Векторюе
	Zeile 24 v. u.	Vector space 1959 anstatt Vector space and Crystal structure, Moskau 1961
S. 502,	Zeile 2 v. o.	. . . Rupprecht, Woodall, Pettit, Growe, Quinn anstatt Alferov, Andreyev, Korol'kov, Tret'yakov, Tuchkevich

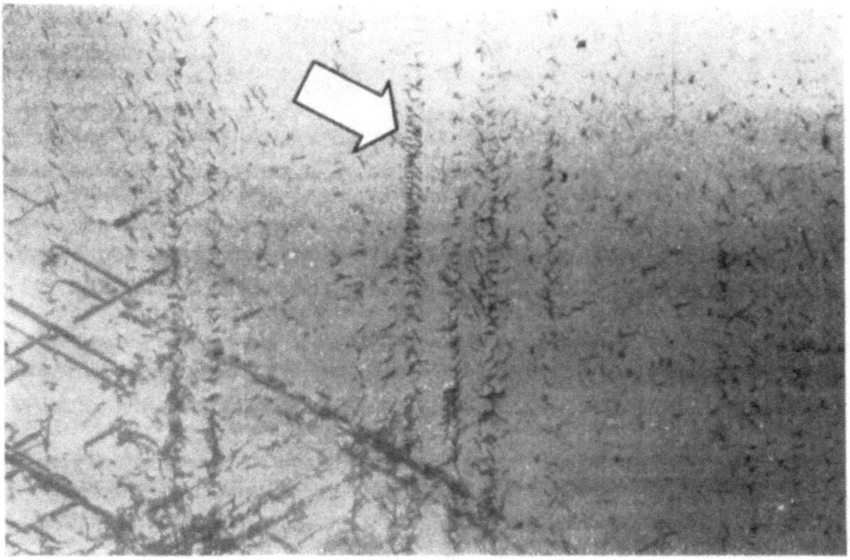

Fig. 1. Piled up dislocations. Substrate thickness 230 μm, layer thickness 230 μm. Lang topogram AgKα_1 beam, (220) reflection

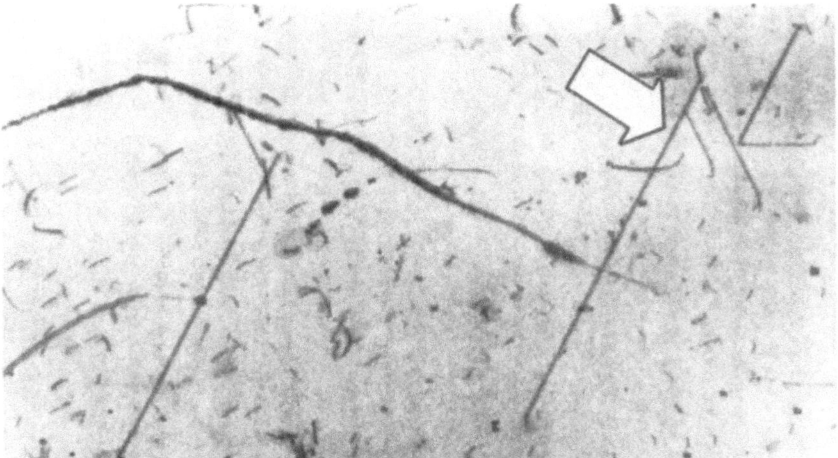

Fig. 2. Large stacking faults in [111] type planes of 20 μm thick Si epitaxial layer of [111] orientation. Lang topogram AgKα_1 beam, (220) reflection

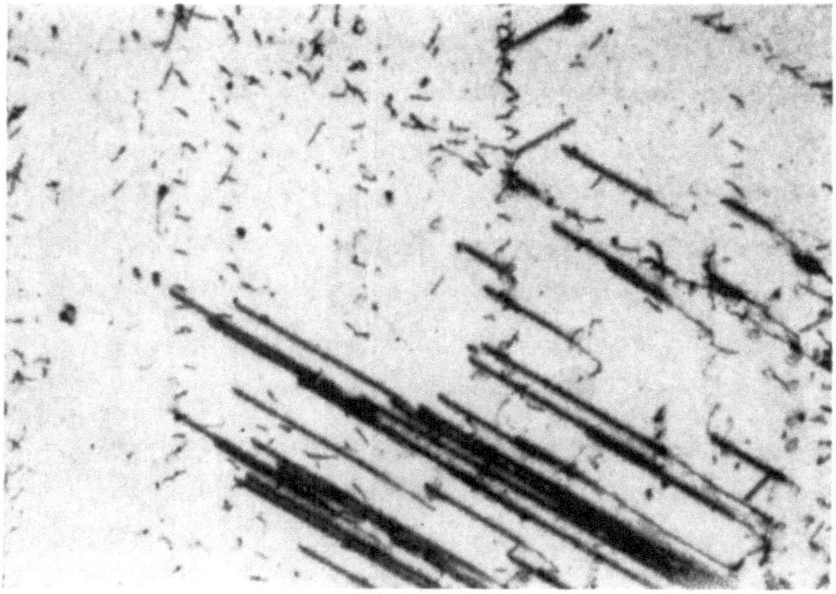

Fig. 3. Accumulated large stacking faults in 20 μm thick Si epitaxial layer of [111] orientation. Lang topogram AgKα_1 beam, (220) reflection

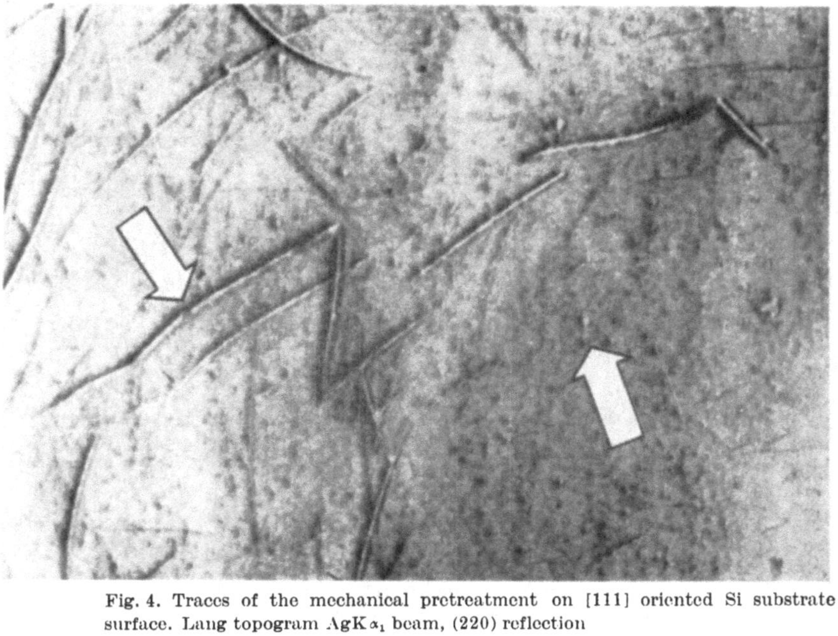

Fig. 4. Traces of the mechanical pretreatment on [111] oriented Si substrate surface. Lang topogram AgKα_1 beam, (220) reflection

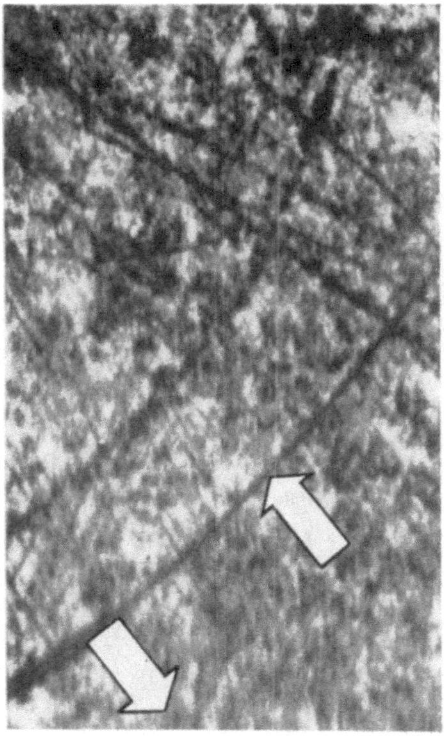

Fig. 5. Traces of strong deformations on [111] oriented Si substrate surface.
Lang topogram AgKα_1 beam, (220) reflection

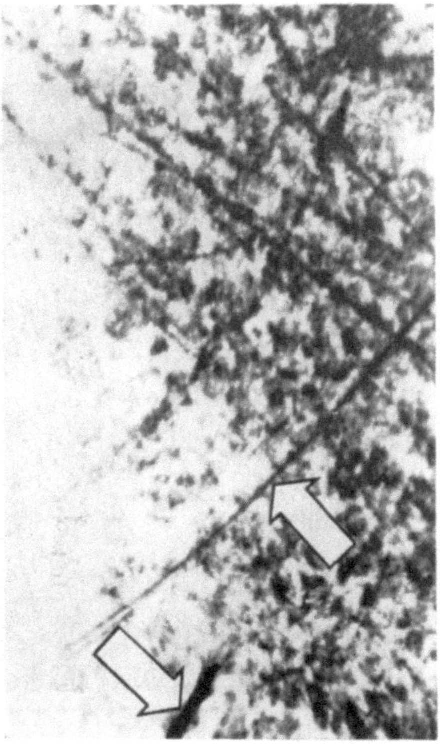

Fig. 6. Substrate as in Figure (5) after grown over by a 30 μm thick Si epitaxial layer.
Lang topogram AgKα_1 beam, (220) reflection

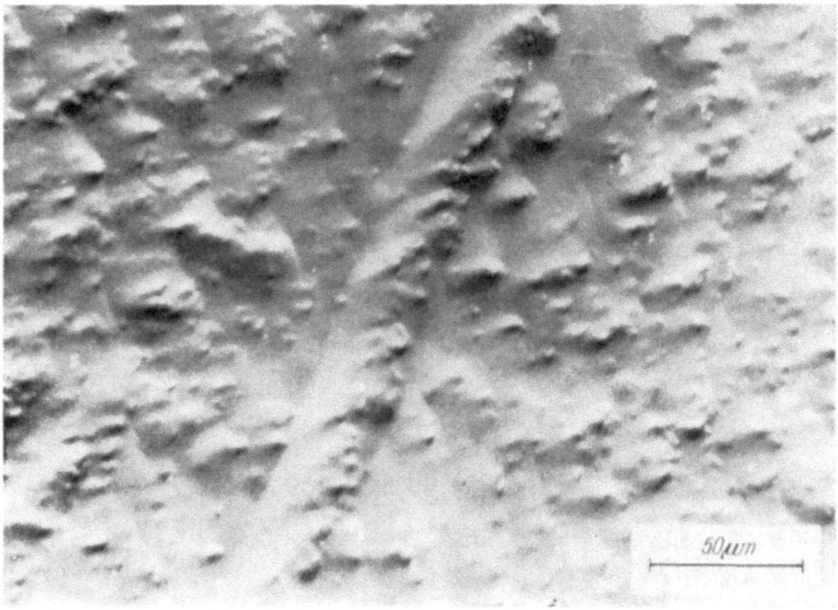

Fig. 7. Interference contrast microscopic pattern of the surface of sample shown on Figure 6

Fig. 8. Laue pattern of defects on a smooth surface of [111] oriented epitaxial Si layer

Kristall und Technik	5	1	1970	K 1 - K 4

C. SZEKELY, L. VARGA

Hungarian Academy of Sciences,
Research Institut for Technical Physics, Budapest

X-Ray Topography of Epitaxially Grown Silicon Layers

The lattice defects of epitaxially grown Si on Si-substrate were investigated by X-ray topography (LANG). Layers of 10-30 μm thickness were grown on different Si-Substrates and have been investigated by means of reflection on the (220) planes. The orientation of the samples was [111]. The development of the defects was investigated by patterns obtained from certain layers of the crystal thinning chemically the epitaxial layer. After each etching process a topographic pattern was taken thus the changes occurring in the substrate-crystal during the epitaxial growth were observable. In certain cases mechanical stresses were developing between the substrate and the layer giving rise to dislocations.

The presence of these dislocations in the substrate was well observable after having removed the epitaxial layer.

Die Gitterstörstellen der auf Si-Träger epitaxial gezüchteten Si-Schichten wurden durch Lang-Topographie untersucht. Schichten von 10-30 μm Dicke wurden auf verschiedenen Si-Trägern gezüchtet und durch Rückstrahlung von (220) Flächen untersucht. Die Orientierung des Präparates war [111]. Die Entwicklung der Störstellen wurde mittels Aufnahmen gewisser Schichten untersucht, die durch chemische Verdünnung der epitaxialen Schicht entstanden. Nach jedem Ätzverfahren wurde ein Topogramm aufgenommen, wodurch die Veränderungen des Trägerkristalls während des epitaxialen Wachstums beobachtet werden konnten. In einigen Fällen entwickelten sich mechanische Spannungen zwischen dem Träger und der aufwachsenden Schicht, wodurch Versetzungen entstanden.

Nach Beseitigung der epitaxialen Schicht konnte die Anwesenheit dieser Versetzungen in dem Träger gut beobachtet werden.

1. Introduction

As SCHWUTTKE proved lattice defects in semiconductor devices are mainly occurring in the course of certain procedures carried out on the base crystal in developing certain semiconductor structures.

These defects should be diminished to the minimum, since in the manufacture they would increase and their elimination would cause sincere difficulties. As known, if the lattice defect density surpasses a certain degree, the stability and scattering resp. of the electric parameters might exceed the prescribed limits.

The lattice defect concentration might considerably increase in the course of epitaxial growth both in the growing layer and in the base crystal. The occurrence of a great number of defects might be eliminated during the manufacture only in knowing its causes and mechanisms of their development.

We intended to observe the defects developing during growth of relatively thin layers of large area also occurring in practice. On the basis of literature (S. MENDELSON 1964, 1965; G.R. BOOKER; J. KÜRTHY, V. STEFAMAI; C. SZÉKELY, L. VARGA) the defect structures to be expected were known. Due to different technical parameters these data did not fully agree with our previous observations. We prefered X-ray topography and interference contrastmicroscopy, both being non-destructive investigation methods (A.R. LANG; G.H. SCHWUTTKE; I. SZÁNTÓ; L. VARGA). For finding the lattice defects in certain cases selective chemical etching and for the examination of local defects the microbeam Laue-technique have been applied.

2. Experimental

Epitaxial growth occurs at a relatively high temperature (1200-1300°C) in a flowing, aggressive chemical medium, in a state of dynamic equilibrium or a state slightly differing from it caused by heating, heat transport, gas-flow, chemical reactions, etc. Both the base material and the layer are influenced by these circumstances and by their interaction. The difference in the lattice parameters caused by the different doping of base and layer might lead to the development of a stress field (sometimes inhomogenious) which at high temperatures is generating dislocations both in the base and in the growing layer.

Also the temperature gradients might cause a stress field. The dislocations in the base crystal are either developing forth in the growing layer or they may curve in the presence of a bending stress field.

In principle dislocations may occur in the interface of two (111) planes of different lattice parameters. It seems to be evident that the lattice defects occurring in the interface should be reduced to a minimum. This might be obtained by decreasing the damage due to the mechanical lapping and polishing and by eliminating the contaminations on the base surface. Considering the afore mentioned the following investigations were carried out:

Undoped layers of a thickness of 10-30μm have been grown onto (111) oriented n-type silicon substrates of a specific resistivity of 0,008 ohm.cm and of a thickness of 0,23 - o,24 mm. The growth was performed decomposing $SiCl_4$ by the aid of H_2 at 1250-1280°C. Before growing "in situ" etching by dry hydrochloric acid gas was applied.

For examining whether the lattice defects were caused by the heat- and etching treatment going with the growth process, blind tests were made at the same temperature but without layer growth.

After having mechanically lapped, polished and chemically etched both surfaces, four slices were examined simultaneously.

According to the interference-contrast microscopical investigations the pretreatment of the surfaces proved to be perfect. Lang topograms were then made of the 8x12 mm^2 medium part of the circular crystal slices by (220) reflection using AgK_{α}, beam of 50 kV and 300μA.

The Lang chamber constructed by the Department of Experimental Physics of the E.I. University of Budapest was fed by a microfocussed beam-source from Hilger. The exposition time was 1 hour pro mm on the Ilford 14 50μm type film.

Having thus fixed the initial stage the blind test was taken and another Lang topogram was made of the same spot of the crystal. After these investigations the homoepitaxial growth took place, followed by interference contrast microscopic and topographic examinations.

A transmission Laue pattern has been taken of the local defects observed by microscope by a 400μm "white" X-ray irradiation at a crystal film distance of 3 cm.

The afore described series of experiments were repeated on further 7 samples, without making blind-tests.

In some experiments the damaged layer on the surface of the base crystal was not fully removed, although according to the interference microscope the surface proved to be perfect. In such cases scratches could be observed but by topography or selected chemical etching and n o t by microscope.

3. Discussion

The lattice defect density of the examined Si crystals was fairly low, it amounted to $1\text{-}10^3/cm^2$. The defect distribution was inhomogeneous even at the initial stage. In certain cases this inhomogeneity reached 3 orders of magnitude. The following three groups of problems have been investigated: a/ piled up dislocations, b/ stacking faults and c/ defects originating from the surface deformation of the base crystal.

a / Piled up dislocations

Piled up dislocations were found in the cleaved (111) and in the inclined (111) slip planes resp. Figure 1 displays a characteristic piled up dislocation in the case of epitaxial growth. The piled up dislocations are nearly parallel. They often are perpendicular to the line of intersection of their own slip plane and the growth plane (Minimum length).

Piled up dislocations may be found in the crystals even in the initial stage, but in these cases the dislocations are hardly identified because of their relatively great distance. The blind test does not give rise to a considerable rearrangement of lattice defect configurations, first of all to that of the piled up dislocations.

In the course of the growth process piled up dislocation series arise but not always. The characteristic patterns of the dislocations piled up in the same slip plane occur on the samples grown on substrates of low defect density. These dislocations appear both in the substrate and in the grown layer, and they remain in the base crystal even after having removed the layer.

Since the effect in the case of a layer of 10μm could be neglected but after the growth of a 30μm thick layer a considerable change could be observed a certain thickness of layer is necessary for obtaining a stress field generating the dislocation sources.

The spatial arrangement of the dislocations proved to be inhomogeneous.

b / Stacking faults

Besides the triangular stacking faults developed by selective chemical etching stacking faults of a few mm length could be observed (Fig.2). These latter accumulated in certain dominions (Fig.3). These stacking faults are most probably limited by partial dislocations, although their structure is not fully known so far. Such defects could be observed only in (111) inclined planes but not in the (111) growth plane.

These defects only developed in the grown layers and could be removed together with the layer by chemical etching. In the case of a thicker layer, the topogram displayed thicker spots as well. The topograms of the defects are strictly straight lines and might form only angles of 60° with each other. Simultaneously with the development of stacking faults not only piled up dislocations, but a stress field of long distance effect could be observed too, which bends the crystal non-uniformly. The contrast difference of the topogram proved its existence.

c / Defects originating from the surface deformation of the base crystal

If the pretreatment of the base crystal was not proper, the remaining deformation traces might be observed even after the epitaxial growth.

Figure 4 displays besides these traces spot-like defects too. Figure 5 shows coarser defects as well. This crystal shows a surface deformation of such a degree that the crystal became bent, thus a considerable part of it was moved off the reflection position. During the growth process, most probably due to the higher temperature this stress field disappeared due to annealing (Fig. 6) however, the imperfect structure has not changed considerably.

Before crystal growth only the defects of the upper layers could be removed, whereas, the defects lying deeper in the crystal would sooner become observable during the growth. A great number of defects are to be found in the interface layers of a sample like this.

Conclusions related to the structure of the layer are difficult to draw from the topogram in this case because of the high concentration of defects in the interface. The inhomogeneous surface of the interference contrast microscope pattern (Fig. 7) may lead to the conclusion of the existence of more imperfect structure.

Besides the afore mentioned ones, other defects are to be observed by the naked eye as well on the smooth and mirror-like surface of the layer. The examination of these imperfect areas showed that some new crystals of grain size of 10-20 μm developed, whose orientation strongly differs from that of the base crystal. Figure 8 displays the Laue pattern of such a crystal.

4. Concluding remarks

In the course of growing a thin homoepitaxial layer dislocations sources affecting the whole Si crystal surface might generate the increase of the lattice defect density both in the base crystal and the growing layer giving rise to well observable piled up dislocation series. Besides the typical triangular stacking faults, such ones of larger size may occur too. After the mechanical treatment of surface the structure of the grown layer is influenced unfavourably by deformations remaining in the deeper parts of the crystal even if the mechanical polishing seemed to be perfect. The grown layer displayed as many or even more defects, as the base crystal.

REFERENCES

BOOKER, G.R.: J. appl. Physics 37, 441 (1966)
KÜRTHY, J., STEFANIAI, V.: IV.Hung.Conf.on X-Ray Diffraction 1968, Esztergom
LANG, A.R.: Acta crystallogr 12, 249 (1959)
MENDELSON, S.: Single Crystal Films, ed.M.H.Francombe-H.Sato, London-New York-Paris 1964
MENDELSON, S.: J. appl. Physics 36, 1315 (1965)
SCHWUTTKE, G.H.: J. appl. Physics 36, 2713 (1965)
SZÁNTÓ, I.: Magyar tud. Akad., mat. fizik. tud. Osztalyanak Közl. R-1 1968
SZÉKELY, C., VARGA, L.: (to be published)
VARGA, L.: (to be published)

(Received April 28, 1969)

Authors' address:

Clara SZÉKELY
L. VARGA
Research Insitute of Technical Physics
Hungarian Academy of Sciences
Budapest IV,
Ujpest 1, Foti ut. 56

VR Ungarn

Kristall und Technik	5	1	1970	K 5 - K 9

Ch. ULLNER

Humboldt-Universität Berlin
Sektion Physik, Bereich Kristallographie

Diffusionsprobleme beim Anfangswachstum dünner Schichten[1)]

Diffusion is a stochastic process. The known differential equation of diffusion theory is valid on assumptions which are not realized always at the beginning of thin film growth. Moreover the stochastic processes of aggregation produce cluster size distribution in which each cluster size has an own probability distribution. In a simple example the probability distribution of adatom concentration is calculated.

Die Diffusion ist ein stochastischer Prozeß. Die bekannte Differentialgleichung der Diffusionstheorie gilt unter Voraussetzungen, die beim Anfangswachstum dünner Schichten nicht in jedem Fall erfüllt sind. Darüberhinaus führen die zufälligen Zusammenlagerungsvorgänge auf Clustergrößenverteilungen, wobei jede einzelne Clustergröße eine eigene Wahrscheinlichkeitsverteilung besitzt. In einem einfachen Beispiel wird die Wahrscheinlichkeitsverteilung der Adatomkonzentation berechnet.

1. Einleitung

Diffusion ist ein Prozeß, der zum Konzentrationsausgleich innerhalb einer einzelnen Phase führt. Die phänomenologische Theorie behandelt den Zusammenhang zwischen der Geschwindigkeit des Diffusionsstromes einer Substanz und dem für diesen Strom verantwortlichen Konzentrationsgradienten. Dabei interessiert die Kinetik der Elementarprozesse überhaupt nicht. Wird jedoch die bekannte partielle Differentialgleichung der Diffusionstheorie in Rechnungen zum Anfangswachstum dünner Schichten auf kristalliner Unterlage verwendet, so ergeben sich bei atomistischer Betrachtung einige Probleme und neue Gesichtspunkte, die hier diskutiert werden sollen.

2. Das Modell

Im allgemeinen wird folgendes Modell angenommen: Die Atome seien "Gitterbausteine". Jede Ortsveränderung ist ein zufälliges Ereignis (im Sinne der Wahrscheinlichkeitstheorie), d.h. sie ist im einzelnen nicht vorausbestimmbar. Lediglich die Wahrscheinlichkeit einer der Realisationsmöglichkeiten läßt sich angeben.

Die Atome mögen die Oberfläche an zufälligen Stellen in zufälligen Augenblicken treffen, sich unmittelbar und vollständig thermisch anpassen (nicht reflektiert werden), sich infolge thermischer Schwankungen in zufälligen Momenten auf einer der benachbarten freien Gitterplätze bewegen oder die Oberfläche wieder verlassen.

---

1) (Vortrag gehalten auf dem Symposium "Kristallisation auf Oberflächen - Grundlagen der Epitaxie" April 1969 Berlin)

3. Brownsche Bewegung eines Einzelatoms

Ist die Konzentration der adsorbierten Einzelatome (Adatome) sehr gering (mögliche Stöße untereinander vernachlässigbar), so werden die Adatome eine Brownsche Bewegung ausführen, da ihre Bewegungen voneinander unabhängig erfolgen und die Sprungwahrscheinlichkeiten in Richtung der positiven und negativen Koordinatenachsen gleich[1] sind. Nur unter diesen Voraussetzungen genügt nach einer Ableitung von EINSTEIN die Aufenthaltswahrscheinlichkeit f(x,t) eines "Brownschen" Teilchens der bekannten Diffusionsgleichung[2]

$$\frac{\partial f}{\partial t} = D \frac{\partial^2 f}{\partial x^2} ,$$

und als deren Lösung, falls sich das Teilchen zur Zeit t=0 mit der Wahrscheinlichkeit Eins im Punkt x=0 befand,

$$f(x,t) = \frac{1}{2\sqrt{\pi Dt}} e^{-\frac{x^2}{4Dt}}$$

Eine häufig benutzte Größe ist die während der Zeitdauer t zurückgelegte mittlere Strecke eines Teilchens, also diejenige Entfernung von der Anfangslage, innerhalb derer sich das Teilchen nach t mit der gleichen Wahrscheinlichkeit befindet wie außerhalb. Diese Entfernung wird näherungsweise durch die Wurzel aus dem mittleren Verschiebungsquadrat bestimmt:

$$\sqrt{\overline{x^2}} = \sqrt{2Dt}$$

Streng genommen gilt diese Formel im Adsorptions-Desorptionsprozeß nur für Einzelatome, von denen vorher bekannt ist, daß sie während t nicht verdampfen. Wird für die Zeit t die mittlere Aufenthaltsdauer τ eines Adatoms auf der Oberfläche eingesetzt, so mag das näherungsweise richtig sein, doch bleibt der Grad der Annäherung im jeweils behandelten Fall unbekannt, zumal bei Steigerung der Adatomskonzentration die Zusammenlagerungsvorgänge berücksichtigt werden müssen.

4. Der stochastische Prozeß

Es ist naheliegend, die hintereinander ablaufenden und aufeinander aufbauenden zufälligen Vorgänge auf der Oberfläche als ganzes zu behandeln. Wird der zufällige Zustand auf der Oberfläche durch den vieldimensionalen Zufallsvektor $\vec{\xi}(t) = \{\xi_i(t)\}$ charakterisiert, so wird die Menge der Zufallsvektoren, die von einem Parameter t abhängen, als stochastischer Prozeß bezeichnet. Die Wahrscheinlichkeitsverteilungsfunktion[3]

$$F(t_o, \varphi ; t, \psi) = P\{\vec{\xi}(t) \quad \psi / \vec{\xi}(t_o) = \varphi\} \tag{1}$$

gibt die Wahrscheinlichkeit dafür an, daß im Zeitraum t die Komponenten des Zufallsvektors $\vec{\xi}(t)$ einen Wert kleiner annehmen als die entsprechenden von ψ, wenn bekannt ist, daß im Zeitpunkt t_o ($t_o < t$) die Gleichung $\vec{\xi}(t_o) = \varphi$ bestand. Diese Verteilungsfunktion dient der vollständigen wahrscheinlichkeitstheoretischen Charakterisierung eines stochastischen Prozesses.

---

1) Sonst gilt die Fokker-Plancksche Gleichung

2) Hier eindimensional, da der zweidimensionale Fall von der Struktur der Kristalloberfläche abhängt

3) Unter Berücksichtigung der Paßverhältnisse

Sollen die Diffusionsprobleme beim Anfangswachstum dünner Schichten umfassend behandelt werden, müßte der Zustand auf der Oberfläche eines Kristalls durch die Zahl der Depositatome über dem i-ten Gitterplatz des Substrats beschrieben werden[1]. Diese Definition des Zustandes entspricht den Simulationsversuchen mittels Rechenautomaten. Nur so können die räumlichen Verhältnisse geklärt werden.

Bei der Betrachtung eines hinreichend homogenen Oberflächenstückes (einschließlich statistisch verteilter Störungen) sind die ortsabhängigen Diffusionsprobleme weniger interessant. Statt dessen ist die Clusterkonzentration $N_i(t)$ als Funktion der Clustergröße - im Folgenden einfach "Größenverteilung" genannt - diejenige Meßgröße, die sicher einmal mit ausreichender Genauigkeit aus Experimenten ermittelt werden kann. Aufgabe theoretischer Arbeiten sollte es daher sein, Clustergrößenverteilungen auf Grund angenommener Modelle zu konstruieren und mit experimentellen Ergebnissen zu vergleichen. Das ist bereits in einigen Fällen getan worden. Was aber noch nicht getan wurde, ist bei einem solchen Vergleich ebenso wichtig: Es müssen die möglichen Schwankungen sowohl der theoretischen wie der experimentellen Kurven wenigstens abgeschätzt werden.

Die Schwankungsbreite der theoretischen Größenverteilung liefert exakt die Wahrscheinlichkeitsverteilungsfunktion des stochastischen Prozesses. Der Zustand auf der Oberfläche sei hier durch die Konzentration der Cluster aus i Atomen charakterisiert. Dann ist die jeweilige Randverteilung von (1)

$$F_i(t_0,\varphi;t,y_i) = \iint \cdots \int \cdot \int \cdots \int F(t_0,\varphi;t,\eta)\, dy_1\, dy_2 \cdots dy_{i-1}\, dy_{i+1} \cdots dy_n$$

die Wahrscheinlichkeitsverteilung der Konzentration von Clustern aus i Atomen. Daraus ergibt sich die Streuung zu

$$D[\xi_i(t)] = (\Delta N_i)^2 = \int y_i^2\, dF_i - (E[\xi_i(t)])^2; \quad E[\xi_i(t)] = \bar{N}_i = \int y_i\, dF_i$$

Die Streuung aller $N_i(t)$ $(i=1,2,\ldots,n)$ bildet die Schwankungsbreite der theoretischen Größenverteilung.

Wie hier deutlich zu sehen ist, sind die mathematischen Schwierigkeiten bei der Lösung dieses vieldimensionalen Problems unüberwindlich. Es muß daher nach Wegen gesucht werden, die Streuung von Größenverteilungen wenigstens abzuschätzen. Im experimentellen Fall ist dies durch eine hinreichend große Menge an Versuchsmaterial möglich.

Im Folgenden wird nach der Theorie der Geburts- und Todesprozesse[2] ein ganz einfaches Beispiel behandelt.

5. Ein einfaches Beispiel

Wir gehen von dem oben angegebenen üblichen Modell aus. Anlagerungsvorgänge bleiben unberücksichtigt. Der stochastische Prozeß sei also auf die Adsorption-Desorption beschränkt. Gesucht ist die Wahrscheinlichkeit $W_k(t)$ dafür, daß sich zur Zeit $t \geq 0$ = k Atome auf der Oberfläche befinden, oder anders ausgedrückt, daß der Prozeß im Zustand k ist.

Der Prozeß kann im Moment $t+\Delta t$ nur dann im Zustand Null sein, wenn er entweder im Moment t in diesem Zustand war und während des Intervalls $[t,t+\Delta t)$ kein neues Atom hinzugekommen ist oder wenn der Prozeß im Augenblick t im Zustand Eins war und im Intervall $[t,t+\Delta t)$ dieses Atom desorbiert ist.

$$W_0(t+\Delta t) = W_0(t)\,(1-P(\Delta t)) + W_1(t) Q_1(\Delta t) \tag{2}$$

---

1) Siehe z.B. GNEDENKO

2) Siehe z.B. FISZ

Die Wahrscheinlichkeit dafür, daß während t,t+Δt) ein Atom hinzukommt, beträgt P(Δt). $Q_i(\Delta t)$ ist die Wahrscheinlichkeit dafür, daß im gleichen Intervall von i Atomen eins desorbiert. Die Wahrscheinlichkeit für das Desorbieren von zwei und mehr Atomen sei von der Ordnung o(Δt). Das gleiche gilt für die Wahrscheinlichkeit, daß während [t,t+Δt)] ein weiteres Atom hinzukommt und ein schon vorhandenes desorbiert.

Im Zustand k (k $\geq$1) kann der Prozeß im Moment t+Δt sein, wenn er entweder schon im Augenblick t im Zustand k war, in dem Intervall t,t+Δt) kein neues Atom hinzugekommen und auch kein altes desorbiert ist oder wenn der Prozeß im Augenblick t im Zustand k-1 war und im Intervall t,t+Δt) ein neues Atom hinzugekommen ist oder wenn der Prozeß im Augenblick t im Zustand k+1 war und im Intervall t,t+Δt) ein Atom desorbiert ist.

$$W_k(t+\Delta t) = W_k(t)(1-P(\Delta t))\,(1-Q_k(\Delta t)) + W_{k-1}(t)P(\Delta t) + W_{k+1}(t)Q_{k+1}(\Delta t) \qquad (3)$$

Die zufälligen Vorgänge des Auftreffens (ausgedrückt durch P(t)) und des Desorbierens (ausgedrückt durch $Q_1(t)$) sind wahrscheinlichkeitstheoretisch nahe verwandt. Betrachten wir eine Zeitskala, auf der in zufälligen Momenten Zeichen gesetzt sind, so entsprechen die Zeichen dem Auftreffen eines Atoms, während die Zeit zwischen den Zeichen Δ die Verteilung der Aufenthaltszeiten der Adatome auf der Oberfläche darstellt. Unter bestimmten Bedingungen[1)], die bei entsprechendem Versuchsaufbau erfüllt sein dürften, ist diese Verteilung eine Exponentialverteilung

$$P\{\Delta < t\} = 1-e^{-\lambda t}$$

mit dem Erwartungswert

$$E[\Delta] = \frac{1}{\lambda} \quad .$$

Daraus ergibt sich wegen der Bedeutung des Mittelwertes und der Ordinarität für hinreichend kleine Zeiten Δt:

$$\begin{aligned} P(\Delta t) &= N{\downarrow}\Delta t + o(\Delta t) \\ Q_1(\Delta t) &= \frac{1}{\tau}\Delta t + o(\Delta t) \end{aligned} \qquad (4)$$

Dabei ist N$\downarrow$ die mittlere Zahl der in der Zeiteinheit auf das betrachtete Oberflächenstück auftreffenden Atome, also die Aufdampfrate. Die mittlere Aufenthaltsdauer ist τ . Bei Vorhandensein von i Atomen, die unabhängig voneinander desorbieren, addieren sich die Wahrscheinlichkeiten:

$$Q_i(\Delta t) = i\frac{1}{\tau}\Delta t + o(\Delta t) \qquad (5)$$

Nach Einsetzen von (4) und (5) in (2) und (3) und Durchführung des Grenzüberganges entsteht das Differentialgleichungssystem:

$$\frac{dW_o(t)}{dt} = -N{\downarrow}\,W_o(t) + \frac{1}{\tau}W_1(t)$$

$$\frac{dW_k(t)}{dt} = -(N + \frac{k}{\tau})W_k(t) + N{\downarrow}\,W_{k-1}(t) + \frac{k+1}{\tau}W_{k+1}(t) \qquad (k=1,2,\ldots)$$

Mit der Anfangsbedingung $W_i(0)=1$ (zur Zeit Null befinden sich mit Sicherheit i Atome auf der Oberfläche) lautet die Lösung:

---

1) Siehe GNEDENKO: Stationär, ohne Nachwirkung, ordinär

$$W_k(t)=\exp\left[-\tau N\downarrow (1-e^{-\frac{t}{\tau}})\right] \sum_{\nu=0}^{\min(i,k)} \binom{i}{\nu} (\tau N\downarrow)^{k-\nu} \frac{e^{-\frac{\nu t}{\tau}} (1-e^{-\frac{t}{\tau}})^{i+k-2}}{(k-\nu)!} \qquad (k=0,\ 1,2,\ldots)$$

Die mittlere Anzahl der Atome, die sich auf der Oberfläche befinden, (der Erwartungswert des Zustandes) beträgt:

$$\bar{k} = E\,[\text{Zustand } k] = \tau N\downarrow (1-e^{-\frac{t}{\tau}}) + i e^{-\frac{t}{\tau}}$$

Die Streuung um diesen Mittelwert ist

$$\overline{k^2} - \bar{k}^2 = D\,[\text{Zustand } k] = (1-e^{-\frac{t}{\tau}})\,(\tau N\downarrow + i e^{-\frac{t}{\tau}}).$$

Für i=0 ist $W_k(t)$ eine Poissonverteilung. Sind die Mittelwerte groß, wird die Poissonverteilung durch eine Normalverteilung gut angenähert, bei der Mittelwert und Streuung gleich sind und zeitlich bis zum stationären Wert anwachsen.

Im stationären Fall beträgt der Mittelwert unabhängig von i $\bar{k} = \tau N\downarrow$. Dieses Ergebnis läßt sich selbstverständlich ebenso aus einfacheren, integralen Betrachtungen ermitteln, doch sollte man bedenken, daß die Behandlung des stochastischen Prozesses exakt die wahren Voraussetzungen herausarbeitet. Außerdem gibt sie uns Aufklärung über die Konzentrationsschwankungen. Die Größe der Streuung von $\tau N\downarrow$ im stationären Fall zeigt, daß unter den Voraussetzungen dieses Beispiels der stochastische Prozeß zu keiner wesentlichen Verschmierung der mittleren Konzentration der Einzelatome führt, falls der beobachtete Flächenausschnitt im Mittel hinreichend viele Atome enthält. Dieses einfache Beispiel kann jedoch keinen Hinweis darauf geben, daß sich die Streuung bei komplizierteren Prozessen, der eigentlichen Keimbildung, ebenso verhält.

6. Schluß

Um das Verhalten der Streuung unter Einbeziehung von Zusammenlagerungsvorgängen zu untersuchen, könnte das bisherige Modell wie folgt erweitert werden: Im Fall vollständiger Abscheidung sei der Zustand durch die Anzahl der Einzelatome und die Anzahl aller Cluster aus zwei und mehr Atomen charakterisiert. Die Wahrscheinlichkeit $W_{ik}(t)$ für das Eintreten des Zustandes i,k genügt ähnlichen Gleichungen wie $W_k(t)$ im obigen Beispiel. Nur sind hier zusätzlich die Wahrscheinlichkeiten dafür enthalten, daß sich während $[t, t+\Delta t)$ zwei der i Atome zu einem Paar zusammenlagern, und dafür, daß sich während $[t, t+\Delta t)$ eines der i Atome an einen der k Cluster anlagert. Die Berechnung dieser beiden Wahrscheinlichkeiten erfordert die Betrachtung spezieller Diffusionsprobleme auf Kristalloberflächen.

LITERATUR

EINSTEIN, A.: Ann. Physik 17, 549 (1905)
FISZ, M.: Wahrscheinlichkeitsrechnung und mathematische Statistik, Berlin 1962
GNEDENKO, B.W.: Lehrbuch der Wahrscheinlichkeitsrechnung, Berlin 1968

(Eingegangen am 30. April 1969)

Anschrift des Verfassers:

Dipl.Phys. Ch. ULLNER
Sektion Physik, Bereich Kristallographie
der Humboldt-Universität
104 Berlin
Invalidenstr. 43

Kristall und Technik	5	1	1970	K 11 - K 14

P. FLÖGEL

Zentralinstitut für Elektronenphysik der Deutschen Akademie der Wissenschaften zu Berlin, Bereich Halbleiteroptik und Sektion Physik der Humboldt-Universität Berlin, Bereich Halbleiteroptik

Zur Kristallzüchtung von Cadmiumsulfid und anderen II-IV-Verbindungen durch Sublimation in verschiedenen Gasgemischen (II)
Chemische Technik zur Gewinnung plättchenförmiger Kristalle des Cadmiumselenids

Es wird der Einfluß verschieden zusammengesetzter N_2-H_2-Transportgase in einer Rohrofenapparatur beschrieben und damit ein ergiebiges Verfahren aufgezeigt.

For growing platelet-shaped CdSe monocrystals the influence of N_2-H_2 carrier gases using a tube furnace is described.

1. Einleitung

Im Anschluß an unsere Untersuchungen zur Gewinnung plättchenförmiger Kristalle des Cadmium- und Zinksulfids (P. FLÖGEL 1968) haben wir die Ausbeute an Cadmiumselenidkristallen bei der Sublimation von CdSe in der gleichen Apparatur (Silitstabrohrofen, 80 cm lang, Sublimationsrohr von Ø 50 mm aus Quarzglas, eingelegtes Kristallisationsrohr Ø 40 mm) mit einigen Transportgasen untersucht. Cadmiumselenid kristallisiert ebenfalls in der Kristallklasse C_{6v} (vgl. z.B. W. KLEBER, I. MIETZ, U. ELSASSER) und kann ähnliche plättchenförmige, mehr oder weniger geriefte, aber schwarz glänzende Einkristalle bilden.

2. Durchführung

Zur Sublimation gelangen jeweils 20 g aus den verdampften Elementen bei 900 °C synthetisiertes, danach in Stickstoff und Wasserstoff ausgeglühtes CdSe (vgl. CdS nach K. ZIMMERMANN) oder das mehrere Stunden lang ausgeglühte "ultrareine" Cadmiumselenid von E. Merck AG. Sein Dampfdruck ist etwa ebenso groß wie der des Sulfids, so daß dieselben Temperaturen um 1000 °C zur Sublimation sich anbieten.

Tabelle 1

Züchtung von CdSe-Einkristallen in verschiedenen Stickstoff-Wasserstoff-Gasgemischen. Züchtungsdauer 3 h, Strömungsgeschwindigkeit 27 l/h, Sublimationstemperatur 1000 °C

Gaszusammensetzung % H_2	% N_2	Verdampfte CdSe-Menge in g	Ausbeute an Plättchen brauchbarer Form
50	50	6,0	15 Stück
25	75	3,75	10 "
8	92	3,0	16 "
4	96	2,6	4 relativ dicke Plättchen
0	100	2,4	0 Stück

Nach Einbringen des Selenids in die bereits auf Sublimationstemperatur erhitzte Zone der Sublimation beobachtet man in der ersten Zeit der Kondensation die Bildung hervorragend wachsender Haarkristalle (Whiskers), der sogenannten "Keime", aus denen sich Plättchen oder Nadeln bilden. Zwei bis vier Stunden später kann die Sublimation durch Beendigung der Ofenheizung zum Erliegen gebracht werden.

3. Ergebnisse

Tabelle 1 gibt die transportierte Menge an CdSe und die Ausbeute an brauchbaren Plättchen für verschiedene Transportgasgemische wieder. Reiner Stickstoff erweist sich als am wenigsten geeignet. Dagegen sind bei den gewählten Parametern Strömungsgeschwindigkeit 12 l/h, Sublimationstemperatur 1000 °C bei Wasserstoffgehalten von 4 bis 50 % durchaus 4 bis 30 Plättchen von 6 bis 250 mm^2 Fläche, 5 bis 25 μm Dicke und nicht zu unebener Oberfläche zu erhalten. Mit steigendem H_2-Gehalt des Transportgases steigt die sublimierte Menge der Substanz an, jedoch nicht so stark wie bei der entsprechenden Sublimation des Cadmiumsulfids. Bei geringerer Konzentration an Wasserstoff entstehen weniger, aber dickere und breitere CdSe-Plättchen als bei größerer Konzentration. Sie sind teils senkrecht, teils schräg zur längeren Ausdehnung der Fläche gestreift (s. Figur 1 a und b). Die röntgenographische Prüfung von sechs Kristallen erwies als Plättchenebene in vier Fällen (11$\bar{2}$0), in zwei Fällen (10$\bar{1}$0); letztere waren schräg gestreift (Fig. 1 b). Die Streifung verläuft parallel zur Hauptachse. Daneben bilden sich reichlich dünne Nadeln bis 40 mm Länge und bis 0,2 mm Breite, mit denen die Plättchen z.T. unvorteilhaft verwachsen sind. In den kälteren Teilen des Quarzglasrohres findet sich nach der Sublimation ein Niederschlag von metallischem Cadmium, in den abziehenden Gasen ist Selenwasserstoff am Geruch zu erkennen.

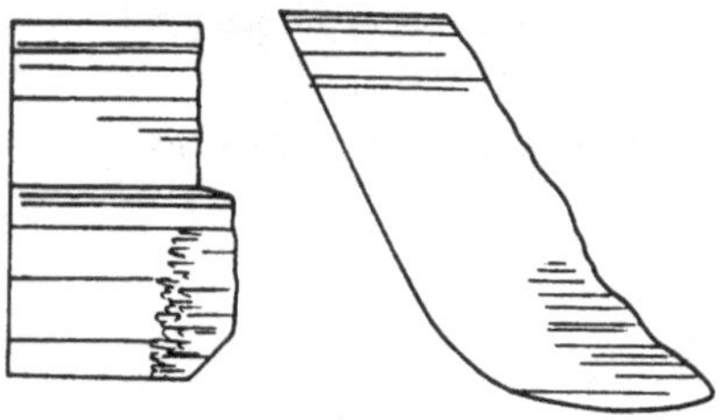

Fig. 1. Plättchenförmige Cadmiumselennidkristalle.
a - senkrecht, b - schräg zur längeren Seite gestreift

4. Diskussion

Wie im Falle des Cadmiumsulfids sind Gemische aus Stickstoff und Wasserstoff auch bei der Sublimation des Cadmiumselenids geeignete Transportgase zur Gewinnung plättchenförmiger (und nadeliger) Einkristalle der gleichen kristallographischen Form. Als besonders vorteilhaft sei hervorgehoben, daß Selenwasserstoff im Transportgas nicht verwendet wird. Der Umstand, daß Wasserstoff hier die Sublimation weniger fördert als beim Sulfid, erklärt sich aus der größeren Dissoziation des H_2Se im untersuchten Temperaturbereich (P. FLÖGEL 1969), so daß der Partialdruck des Selens im Gleichgewicht

$$CdSe_{fest} + H_2 = Cd + 1/2\ Se_2 + H_2 = Cd + H_2Se$$

durch Wasserstoff nicht so stark vermindert wird, wie der des Schwefels bei der entsprechenden Reaktion mit dem Sulfid. Daraus erklärt sich auch die Möglichkeit, das Stickstoff-Wasserstoff-Verhältnis im Gas ohne Schaden für die Ausbeute in weiteren Grenzen variieren zu können als bei der entsprechenden Kristallzüchtung des Sulfids.

5. Besondere Varianten

Ebenfalls in Analogie zur Gewinnung von CdS-Plättchen wachsen auch in einem Strom reinen Stickstoffs als Transportgas hinreichend glatte, große, breite Plättchen, wenn zuvor ihre Keimbildung in einem Gemisch von Stickstoff und Wasserstoff stattgefunden hat.

Anwendung höherer Temperaturen zur Sublimation als 1000 °C ergibt im Sublimat bei Temperaturen für das Wachstum von 940 und mehr °C vorzugsweise Nadeln, längs der hexagonalen Achse gestreckt, bis zu einigen Zehntel mm Dicke und mehreren mm Länge, deren freies Ende meist hohl ausgebildet ist. Nicht selten sind sie mit einem Plättchen unvorteilhaft verwachsen. Die Ausbeute an gut gewachsenen Plättchen sinkt dagegen. Will man sich zur Gewinnung besonders dicker Plättchen höherer Temperaturen bedienen, so ist zuvor eine "Keimbildung" für Plättchen bei hinreichend tiefer Temperatur angeraten. Will man sich zur Gewinnung besonders dicker Plättchen höherer Temperaturen bedienen, so ist zuvor eine "Keimbildung" für Plättchen bei hinreichend tiefer Temperatur angeraten. Bei Temperaturen über 1000 °C macht sich in Wasserstoff enthaltenden Gasen der Transport von Siliciumverbindungen aus dem Quarzglas durch matte grauweiße Ablagerungen auf den sonst glänzenden Kristallen bemerkbar. Er kann vermieden werden, wenn man das Transportgas nach erfolgter sichtbarer "Keimbildung" vor dem Eintritt in den Ofen noch durch eine vorher vom Transportgas schon ausgespülte Waschflasche mit Wasser leitet.

Bei Anwendung tieferer Temperaturen als 1000 °C zur Sublimation nimmt die Bildung von Plättchen im Sublimat ab, ist aber bei ca. 825 °C noch zu beobachten (Wachstumstemperaturen um 750 °C, Transportgas 20 l N_2 + 10 l H_2 / h). Die erfolgreiche Keimbildung wird dabei erst 0,5 bis 1,5 h nach Beginn der Sublimation sichtbar. Zweckmäßig erhöht man die Ofentemperatur danach etwas.

Diese Bedingungen sind günstig sowohl für die Gewinnung der besonders dünnen, in Interferenzfarben schillernden Plättchen, die in neuester Zeit wissenschaftliche Beachtung gefunden haben (H. LANGE, E. GUTSCHE), als auch für eine Ausbildung von Plättchen ohne Verwachsungen mit Nadeln.

6. Versuche mit ZnSe

Versuche, plättchenförmige Kristalle des Zinkselenids in derselben Apparatur bei Temperaturen von 1040 bis 1200 °C für die Sublimation und bei verschiedenen Verhältnissen von Stickstoff und Wasserstoff im Gasstrom zu erhalten, waren nicht erfolgreich. Das Kondensat zwischen 900 und 1140 °C bestand jedes Mal nur aus kleineren gelblichen kubischen Kristallen, kenntlich an optischer Isotropie und an (111)-Flächen mit Mustern aus parallelen gleichseitigen Dreiecken und aus einigen dendritisch verzweigten, hellgelben, nicht einheitlich doppelbrechenden Nadelm vom Durchmesser unter 0,1 mm.

Dem Leiter des Bereiches, Herrn Prof. Dr. E. Gutsche, und Herrn Dr. Wolfgang Borchardt danke ich für ihre ständige Förderung, Herrn Dr. Werner Möhling für die röntgenographische Prüfung einiger CdSe-Plättchen, meinen Mitarbeitern für erfolgreiche Zusammenarbeit.

LITERATUR

FLÖGEL, P.: Kristall und Technik 3, 161-175 (1968) Z. anorg.u.allg. Chemie 1969 (im Druck)
KLEBER, W., MIETZ, I., ELSASSER, U.: Kristall und Technik 2, 327-338 (1968)
LANGE, H., GUTSCHE, E.: Physica status solidi Berlin 32, 293 (1969)
ZIMMERMANN, K.: Dissertation, Martin-Luther-Universität Halle/Wittenberg 1961

(Eingegangen am 15. Mai 1969)

Anschrift des Verfassers:

Dr. Peter FLÖGEL
Sektion Physik der Humboldt-Universität
102 Berlin, Neue Schönhauser Str. 20

Kristall und Technik	5	1	1970	K 15 - K 17

B.M. BULAKH, G.S. PEKAR

Institut of Semiconductors, Academy of Sciences
of the Ukrainian SSR, Kiev, USSR

Heat Treatment of CdS Large Single Crystals under Controlled Conditions

Recently the need for large volume single crystals of CdS with different parameters has greatly increased. Large CdS crystals are grown from the vapour phase or from the melt. At the same time the as-grown crystals are usually non-photosensitive with a low dark resistivity. It is considered to result from the excess Cd in the crystals (WOODBURY; HANDELMAN, THOMAS; UCHIDA; CLARK, WOODS 1966). It is possible to change crystal properties by varying the vapour phase composition during crystal growing (FOCHS, GEORGE, AUGUSTUS; CLARK, WOODS 1968), but it was shown to affect strongly on the crystal growth (BULAKH, MIZETSKAYA; HEMMAT, WEINSTEIN).

The more convenient way for receiving samples with different parameters is the crystal growing under optimal conditions followed by heat treatment of the samples. Recently the effect of the treatment in sealed ampoules (in vacuum and under S or Cd vapour pressure) on crystal properties was investigated (WOODBURY; HANDELMAN, THOMAS; UCHIDA). The unreproducible results after the first treatment in evacuated sealed ampoules were observed sometimes; only secondary heating in other ampoules produced the expected effects (HANDELMAN, THOMAS).

We investigated the high-temperature treatment effects on the photosensitivity and dark resistivity of the large volume CdS single crystals. The low resistivity crystals (0.01 ÷ 1.0 ohm·cm) were grown from the vapour phase with the method which is similar to that described by PIPER and POLICH. The different methods of heating were used - in vacuum, in inert gas flow, in sealed ampoules under sulphur vapour pressure, in the atmosphere of inert gas and sulphur under controlled sulphur pressure. The samples were of nearly cubic shape with the weight varying from 0.5 to 25.0g. A conclusion was drawn that during heating of the large crystals the secondary chemical reactions particularly with the participation of oxygen began to play the main part. The following could represent such a series of reactions.

$$Cd + 1/2\,O_2 = CdO \tag{1}$$

$$CdS + 3/2\,O_2 = CdO + SO_2$$

$$CdS + O_2 = Cd + SO_2$$

The reaction (1) proceeds already when the oxygen pressure is $2.12 \cdot 10^{-12}$ torr at 800°C (CHIZHIKOV et al.)

At high temperatures Cd0 can interact with CdS:

$$CdS + 2CdO = 3Cd + SO_2.$$

As RIEHL and SIZMANN showed the SO_2 partial pressure in the CdS+Cd system is 144 torr at 900°C and 320 torr at 1000°C.

In the case of high-temperature crystals treatment under S vapour pressure interaction between CdO and S proceeds:

$$2CdO + 3S = 2CdS + SO_2.$$

So, if heating in the sealed ampoules is used the reactions in which gaseous products form can proceed not to the end. That is why the results of such heating depend not only on temperature of treatment, time, sulphur pressure but on ampoule volume, crystal size and quantity of the excess Cd in the crystals. This question was discussed in details previously (BULAKH, PEKAR).

The method of high-temperature heat treatment of CdS crystals in the inert gas flow under controlled sulphur vapour pressure has been developed. It gives the posibility to avoid the negative influence of secondary chemical reactions.

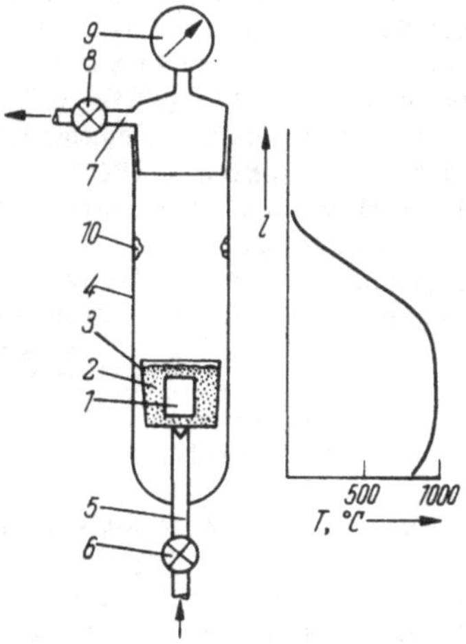

Fig. 1. Arrangement for heating CdS crystals in the inert gas flow under controlled sulphur vapour pressure

The arrangement for heating is shown schematically in Figure 1. The specimen (1) together with purified CdS powder (2) and some sulphur were placed in the quartz crucible (3). The crucible was then inserted in the quartz tube (4) with two orifices (5), the orifice (7) was joined to a pumping out system. It was possible to change pressure in the tube by regulating the gas supplying and the ratio of pumping out with two valves (6) and (8). Pressure was controlled with a manometer (9).

The tube was set inside the furnace with a temperature profile shown in Figure 1. Sulphur evaporated, then condensed in the cold zone (10) and flowed down to the zone of evaporation. So, it was possible to control the sulphur vapour pressure directly by regulating the pressure in the cold zone (VAN DOORN). The inert gas flow removed ''O_2 from the system which promoted chemical reactions to proceed to the end. The CdS powder prevented crystal sublimation which usually is very considerable during high-temperature heating in the gas flow.

The samples after heating were optically uniform. Crystals changed in colour from brown (due to excess Cd (WOODBURY; CLARK, WOODS 1966)) to orange-yellow. If the heating time was long enough the parameters of samples were independent on their size. Crystals with different photosensitivity and dark resistivity were received depending on sulphur pressure during heating. The ratio of dark resistivity (ρ_d) to the resistivity under light (ρ_l) as a function of sulphur vapour pressure during heat treatment is shown in Figure 2.

The authors wish to thank Dr. M.K. Sheinkman for helpful discussions.

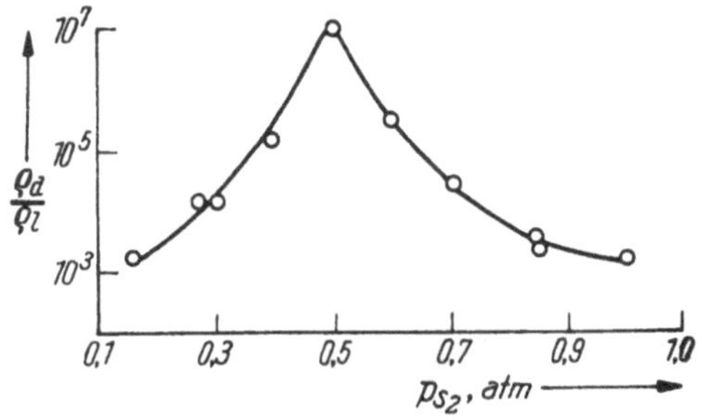

Fig. 2. The ratio of crystals dark resistivity (ρ_d) to resistivity under light 10^4 luxes (ρ_l) as a function of sulphur vapour pressure during heat treatment

REFERENCES

BULAKH, B.M., MIZETSKAYA, I.B.: Ukr.Fiz.Zh., 7, 1125 (1962)
BULAKH, B.M., PEKAR, G.S.: Zh.Neorgan.Materiali (in print)
CHIZHIKOV, D.M., FRENTS, G.S., TRATSEVITSKAYA, B.Y.: Izv.Akad.Nauk SSSR 12, 1815 (1950)
CLARK, L., WOODS, J.: Brit.J.appl.Physic 17, 319 (1966)
CLARK, L., WOODS, J.: J.Cryst.Growth 3, 4, 127 (1968)
FOCHS, P.D., GEORGE, W.; AUGUSTUS, P.D.: J.Cryst.Growth 3, 4, 122 (1968)
HANDELMAN, E.T., THOMAS, D.G.: J.Physics and Chem.Sol. 26, 1261 (1965)
HEMMAT, N., WEINSTEIN, M.: J.Electrochem Soc. 114, 851 (1967)
PIPER, W.W., POLICH, S.I.: J.appl.Physics 32, 1278 (1961)
RIEHL, N., SIZMANN, R.: Z.Naturforsch. B14a, 394 (1959)
UCHIDA, I.: J.Physic.Soc.Japan 22, 771 (1967)
VAN DOORN, C.Z.: Rev.Sci.Instr. 32, 755 (1961)
WOODBURY, H.H.: Physic.Rev. 134, 492 (1964)

(Received June 12, 1969)

Authors' address:

B.M. BULAKH, G.S. PEKAR
Institute of Semiconductors,
Academy of Sciences of the Ukrainian SSR,
B.Kitaevskaya 115
Kiev-28,
USSR

Kristall und Technik	5	1	1970	K 19 - K 22

Jiri KVAPIL

Monokrystaly, Turnov, CSSR

Der Einfluß der Temperatursymmetrie des Verneuilofens auf die Qualität der Korundeinkristalle

In Verneuilöfen mit eliptischem Querschnitt der Isothermen senkrecht zur Wachstumsachse wurden Korundbirnen gezüchtet. Die gezüchteten Einkristalle wurden mit einer Geschwindigkeit von 250°C/h auf eine Temperatur von 1850°C gebracht. Die Anzahl der beim Erhitzen gesprungenen Einkristalle wurde zum Winkel ϱ der kristallographische c-Achse mit der Wachstumsachse als auch zum Winkel α in Beziehung gesetzt. Der Winkel α wird durch zwei Ebenen gebildet. Die erste Ebene ist durch die Wachstumsachse und durch die längere Achse des eliptischen Querschnitts der Isotherme bestimmt, die zweite durch die Wachstumsachse und die kristallographische c-Achse. Die höchste Stabilität zeigen stets die Einkristalle mit dem Winkel $\alpha \doteq 45°$, die niedrigsten die mit $\alpha \doteq 90°$ oder $\alpha \doteq 0°$.

Corundum boules were grown in Verneuil furnaces with eliptical cross sections of the isotherms perpendicular to the growth axis. The as grown single crystals were heated with a rate of 250°C/h to a temperature of 1850°C. The number of single crystals which split during heating was evaluated with regard to the angle ϱ of the c-axis with the growth axis as well as with regard to the angle α. The angle α is enclosed by two planes. The first plane ist established by the growth axis and by the longer axis of the eliptical cross section of the isothermes. The second plane ist established by the crystallographic c-axis and by the growth axis. The greatest stability is always observed with single crystals with the angle $\alpha \doteq 45°$, the lowest with those having $\alpha \doteq 90°$ or $\alpha \doteq 0°$.

1. Einleitung

Die Mehrzahl der Arbeiten, welche die Züchtung der Korundeinkristalle durch die Verneuilmethode beschrieben, befaßt sich nur mit dem Einfluß der Axialkomponente des Temperaturgradienten (ADAMSKI; HUTCHESON; ORDWAY, MELLER; REISS). Der Radialkomponenten des Temperaturgradienten bzw. der Form der Isotherme im Hohlraum des Verneuilofens wurde nur geringe Aufmerksamkeit gewidmet. Es wird angenommen, daß die Isothermen rotationssymmetrische Flächen sind, wobei die Wachstumsachse gleichzeitig auch die Rotationsachse ist. Die rotationssymmetrische Form der Isotherme wird in der Praxis nur selten eingehalten. Die Isothermen werden besonders an der Berührungsstelle der keramischen Formsteine als auch am Betrachtungsfenster deformiert. In diesen Fällen ist die Qualität der gezüchteten Kristalle auch von der Orientierung der Kristallachsen in Bezug auf die Lage der Isothermen abhängig.

2. Experimenteller Teil

In Verneuilöfen, die aus zwei halbrunden Formsteinen bestehen (Fig 1.b), wurden Korundbirnen auf die früher beschriebene Weise (KVAPIL) gezüchtet, und zwar mit der Wachstums-

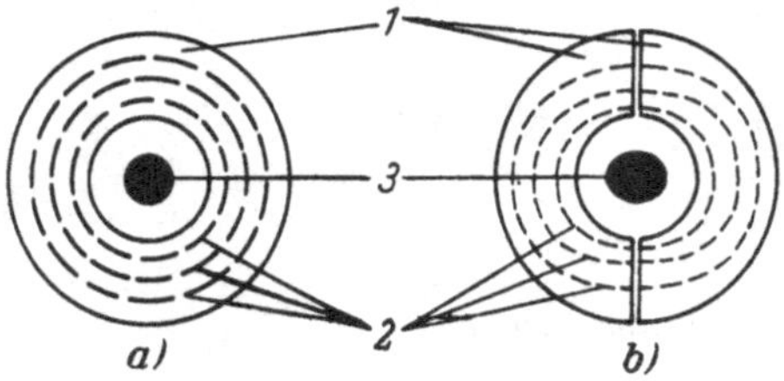

Fig. 1. Zerlegung der Isothermen im Querschnitt des Verneuilofens. 1 - Keramische Formsteine, 2 - Isothermen, 3 - Korundbirne.
a) Ofen, aus einem Stück hergestellt
b) Ofen, aus zwei Formsteinen hergestellt

orientierung gegen die c-Achse: $58^o < \rho < 73^o$, und mit veränderlicher Orientierung der kristallographischen a-Achse. Der Keimkristall hatte stets denselben Chromgehalt wie die auf ihm gezüchtete Birne. Bei jedem Einkristall wurde der Winkel α der Ebenen π_1 und π_2 bestimmt. Die Ebene π_1 ist durch die Wachstumsachse und durch die kristallographische c-Achse bestimmt, die Ebene π_2 durch die Wachstumsachse und durch die längere Achse des Querschnitts der Isothermen (senkrecht zur Wachstumsachse.

Der Verlauf der Isothermen im Züchtungsofen bzw. an dessen Formsteinen wurde mittels Metall-Farboxiden sichtbar gemacht (BARTA, ŽEMLIČKA, KMENT). Diese wurden der keramischen Mischung für die Ausmauerung beigegeben.

Die erhaltenen Kristalle wurden nach der Größe des Winkels ρ in drei Gruppen geteilt: $58^o < \rho < 63^o$; $63^o < \rho < 68^o$; $68^o < \rho < 73^o$ und jede Gruppe in sechs Untergruppen je nach der Größe des Winkels α : $0^o < \alpha < 15^o$; $15^o < \alpha < 30^o$; $30^o < \alpha < 45^o$; $45^o < \alpha < 60^o$; $60^o < \alpha < 75^o$; $75^o < \alpha < 90^o$.
Die gleiche Menge an Einkristallen aus jeder dieser achtzehn Untergruppen wurde weiter erhitzt, wie früher beschrieben (KVAPIL). Nach der Erhitzung wurde die Menge der gesprungenen Kristalle in Abhängigkeit vom Winkel ρ und α bestimmt. In der Tabelle 1 beträgt der Gesamtanteil der gesprungenen Einkristalle in jeder von den drei Gruppen 100 %. Die Orientierung ist im Keimgebiet gemessen.

Tabelle 1

$58^o < \rho < 63^o$		$63^o < \rho < 68^o$		$68^o < \rho < 73^o$	
$\alpha^{(o)}$	gespaltene Birnen (%)	$\alpha^{(o)}$	gespaltene Birnen (%)	$\alpha^{(o)}$	gespaltene Birnen (%)
0-15	22	0-15	25	0-15	29
15-30	18	15-30	20	15-30	20
30-45	14	30-45	12	30-45	7
45-60	12	45-60	8	45-60	5
60-75	16	60-75	15	60-75	16
75-90	18	75-90	20	75-90	23

Die Einordnung der Kristalle nach der Orientierung der kristallographischen a-Achsen und nach dem Chromgehalt wurde nicht vorgenommen. Orientierende Untersuchungen zeigten jedoch, daß dieser Faktor die Abhängigkeiten, die in der Tabelle 1 angegeben sind, nicht beeinflussen.

3. Diskussion

Die Form des Kristallquerschnittes wird vor allem durch die Form der Isothermen bestimmt (NEUHAUS, BRENNER). Dies bedeutet, daß sich die Korundbirne nicht hinter die

Schmelzpunktisotherme des Korunds ausbreiten kann. Im Fall, daß der Querschnitt der Isotherme elliptisch ist, wächst auch der Kristall mit einem eliptischen Querschnitt (Fig. 1 b). Es ist auch bekannt, daß sich die Korundbirnen bei der Wachstumsorientierung $\varrho \approx 90^o$ in der Richtung senkrecht zur kristallographischen c-Achse strecken und daß sich die (0001) Flächen entwickeln. (RUDNICKAJA).

Im Fall $\alpha \doteq 0^o$ ist die Entwicklung des Kristalls senkrecht zur Richtung der kristallographischen c-Achse durch die Schmelzisotherme begrenzt. Wie die Messung der Orientierung längs der Wachstumsachse gezeigt hat, reagiert der wachsende Kristall auf die angeführten Bedingungen auf eine solche Weise, daß sich beim Wachsen der Winkel α (nebst der üblichen Zunahme von ϱ) vergrößert. Bei einer Länge von 70 mm wurde für $0^o < \alpha < 15^o$ (für den Keim) eine Erhöhung von α im Durchschnitt um 5° festgestellt, für $30^o < \alpha < 45^o$ um 1 - 2° und für $75^o < \alpha < 90^o$ hat sich der Winkel α nicht geändert. Die Verdrehung des Kristallgitters ist durch die Erhöhung des Grades der Störung bedingt (Fig. 2 a), und das hat eine niedrige Temperaturstabilität bei der Aufheizung als auch bei der Züchtung selbst zur Folge. Im Fall $\alpha \doteq 90^o$ entspricht die günstigste Richtung für den Stoffansatz des Kristalls der Richtung der Verdreihung des Querschnittes der Isothermen, so daß der mit Flächen versehene Einkristall ohne Verdrehung der Fläche π_1 wachsen kann, ohne den hohen Störungsgrad durch den Mosaikbau zu haben. (Fig. 2 b). Die Biegungsmomente, die gewöhnlich die Spaltung der Korundbirnen verursachen (INDENBOM, TOMILOVSKIJ), sind dagegen in diesem Fall sehr stark (der Kraftarm ist länger), und deshalb kommt es leicht zur Spaltung der Kristalle. Die höchste Stabilität erweisen die Einkristalle bei einer Orientierung $\alpha \doteq 45^o$. Mit der Verkleinerung des Winkels ϱ sinkt der Einfluß der Wachstumsanisotropie auf den Querschnitt des Kristalls und dadurch auch die Abhängigkeit der Stabilität der Kristalle vom Winkel α (Tab. 1).

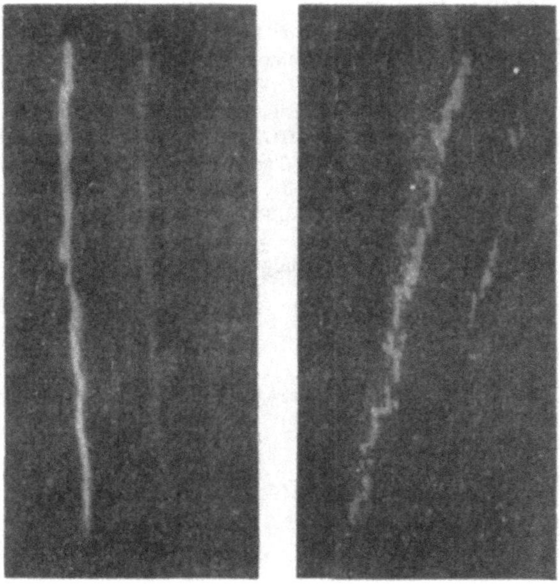

Fig. 2. Störungsgrad der Korundbirnen, röntgenographisch mit dem oszillierenden Spektrometer (AULEYTNER) längs der Wachstumsachse gemessen. Benutzte Reflexion (0006) - Cu $K\alpha$ - und Cu $K\beta$ -Linien. Die Verschiebung der Linien um 1 mm entspricht der Drehung der Ebene π_1 um $0{,}2^o$. Orientierung des Keimkristalle $\varrho = 90^o$, a) $\alpha = 10^o$, b) $\alpha = 85^o$.

Die beschriebene Abhängigkeit hat bei der Züchtung der Korundbirnen eine relativ kleine Bedeutung, weil die Deformation der Isothermen durch günstige Konstruktion des Ofens, eventuell durch Rotation des Kristalls unterdrückt werden kann. Eine außerordentliche Bedeutung hat sie für die von Flächen begrenzten Einkristalle, die auf den nicht bewegten oder auf den geradlinig hin und her bewegten Keimkristallen gezüchtet wurden (BARTA; DJEVAHIRDJIAN; DROST, KEBLER, DUTCHLESS). In diesem Falle entspricht der Querschnitt der Flächenisothermen der extremen Verlängerung des Querschnitts des Einkristalls. Aus diesem Grund beeinflußt vor allem der Winkel α die Stabilität dieser Kristalle.

4. Zusammenfassung

Die Stabilität von Korundeinkristallen, die nach der Verneuilmethode gezüchtet wurden, ist von der Orientierung der kristallographischen c-Achse gegen die Achsen des Querschnitts der Isothermen beim Wachstum abhängig. Die größte Stabilität haben Kristalle, die so gezüchtet wurden, daß die Projektion der kristallographischen c-Achse auf die Fläche des Querschnitts der Isotherme (senkrecht zur Wachstumsachse) mit den Achsen dieses Querschnitts einen Winkel von 45° einschließt.

LITERATUR

(Eingegangen am 12. Juni 1969)

ADAMSKI, J.A.: J. appl. Physics 36, 1784 (1965)
AULEYTNER, J.: X-ray Methods in The Study of Defects in Single Crystals, PWN Warszawa 1964, S. 86
BARTA, Č.: Rost kristallov 6, 181 (1965)
BARTA, Č., ŽEMLIČKA, J., KMENT, V.: Sbornik referatu IV. konference o monokrystalech, Turnov 1962, S. 27
DJEVAHIRDJIAN, V.: Schweiz. Pat. 354 428 (1961)
DROST, W., KEBLER, R.W., DUTCHLESS, E.E.: USA Pat. 3 077 752, 1961
HUTCHESON, R.L.: IRE International Conv. Record Pt 3, 10, 15 (1962)
INDENBOM, V.L., TOMILOVSKIJ, G.E.: Kristallografija 3, 593 (1958)
KVAPIL, J.: Kristall u. Technik 4, 237 (1969)
NEUHAUS, A., BRENNER, K.: Chem. Ing. Technik 27, 320 (1955)
ORDWAY, F., MELLER, P.R.: Ceram. Bull. 43, 253 (1964)
REISS, F.A.: Appl.Optics 5, 1902 (1966)
RUDNICKAJA, E.S.: Trudy IKAN SSSR 8, 13 (1953)

(Eingegangen am 12. Juni 1969)

Anschrift des Verfassers:

Ing. Jiří KVAPIL
Monokrystaly
Turnov
Leninova 175
ČSSR

Kristall und Technik	5	1	1970	K 23

J. LIEBERTZ

Philips Zentrallaboratorium GmbH, Laboratorium Aachen

Hydrothermale Kristallisation von $SrSO_4$, $BaSO_4$ und $PbSO_4$ aus wäßrigen H_2SO_4-Lösungen

Die in reinem Wasser bei Raumtemperatur nahezu unlöslichen Sulfate von Strontium, Barium und Blei lösen sich merklich in heißer konzentrierter Schwefelsäure. Es erschien daher von Interesse, die Kristallisation unter hydrothermalen Bedingungen aus H_2O-H_2SO_4-Lösungen wechselnder Zusammensetzung zu verfolgen.

Die Versuche wurden bei 350°C über einen Zeitraum von jeweils 4 Wochen durchgeführt. Die Konzentration der Schwefelsäure variierte in Stufen von 10 % von 20 bis 100 Gew.-%. In Anlehnung an die Technik von RAU und RABENAU dienten als Gefäße Quarzglasampullen (Inhalt ca. 5 ml), die mit 1 g gefälltem Sulfat und der Lösung zu 80 % gefüllt und dann abgeschmolzen wurden. Je drei Ampullen paßten nebeneinander in einen Autoklaven, dessen freies Volumen zur Druckkompensation mit einer geeigneten Menge Trockeneis (CO_2) gefüllt wurde.

Die besten Ergebnisse wurden in allen drei Fällen bei Säuregehalten von 70 bis 90 % erzielt. Hier waren vorwiegend im oberen, kälteren Teil der Ampulle Kristalle der Größe um 2 mm gewachsen, verglichen mit ca. 0,2 mm bei 20 %-iger Säure. Die $SrSO_4$- und $BaSO_4$-Kristalle waren meist gut ausgebildet und zeigten prismatischen Habitus mit den Flächen (011), (001) und (210). Beim $SrSO_4$ fanden sich auch (001)-Tafeln mit den weiteren Flächen (011), (101) und (210). Beim $PbSO_4$ traten neben schlecht ausgebildeten prismatischen Kristallen, deren Flächen sich nicht vermessen ließen, auch Hohlformen und dünne, bis 5 mm lange Nadeln nach [010] auf. Da letztere vermutlich nicht durch Transport, sondern erst im Zuge der Abkühlung bei Versuchsende gewachsen sind, kann auf eine unter den herrschenden Bedingungen beträchtlich erhöhte Löslichkeit geschlossen werden.

LITERATUR

RAU, H., RABENAU, A.: Mat. Res. Bull. 2, 609 (1967)

(Eingegangen am 11. August 1969)

Anschrift des Verfassers:

Dr. Josef LIEBERTZ
Philips Zentrallaboratorium GmbH, Laboratorium Aachen
51 Aachen, Weishausstr.
Jetzt Martinswerk GmbH, 515 Bergheim/Erft

Kolibri 360 GmbH
Ferdinand-Jühlke-Straße 7
99095 Erfurt, Deutschland
produktsicherheit@kolibri360.de